UNDERSTANDING YOUR ZX81 ROM

By

Dr Ian Logan

First Published in 1981 by Melbourne House

This Remastered Edition
Published by
Acorn Books
acornbooks.uk

This book is a page-by-page reproduction of the original 1981 edition as published by Melbourne House. The entirety of the book is presented with no changes, corrections nor updates to the original text; therefore no guarantee is offered as to the accuracy of the information within.

==

LATE NEWS EXTRA ** LATE NEWS EXTRA ** LATE NEWS EXTRA ** LATE NEWS E

==

SINCLAIR DEVELOPS A 'NEW' 8K ROM!

The original 8K ROM issued by Sinclair Research in March 1981 has unfortunately been found to contain several errors. The major problem is errors in calculations:
e.g.
PRINT 0.25 ** 2 gives 3.1423844, which is rather a long way from the correct answer.

As well, the line POKE 16437, 255 has to be added for safety after every use of the PAUSE command to prevent the 'white-out' effect.

These failures in the program of the 8K ROM have resulted in the issue of a 'new' 8K ROM by Sinclair Research.

This new ROM can be identified by obtaining the following results:

- PRINT 0.25 ** 2 gives the correct answer 0.0625
- PRINT PEEK 54 gives 136

(Both tests are necessary as there are just a few ZX 81s that contain a hardware add-on to make them count properly, but they do not contain the 'new' ROM.)

The differences between the 'new' and the 'old' ROMs are:

i) The workspace is cleared in the INPUT command routine (a house-keeping error)
ii) FRAMES-high, 16437, is loaded with 255 on every return from PAUSE. (a very good idea.)
iii) A change has been made in the VARIABLE routine at 102F Hex. (the reason for this is not known as yet)
iv) And most importantly the three troublesome 'extra'bytes at 1733 - 1735 Hex. that lead to the arithmetic error are simply deleted.

Although it is as yet unconfirmed by Sinclair, it would appear that the hardware add-on that corrects the arithmetic on some of the 'old' 8K ROMs works by nullifying the effect of the 'three bytes'. Probably when the location 1735 is addressed the instruction fetched is 'LD H,A' but the instruction passed to the Z80 is 'DAA'.

Readers of this book that are using machines fitted with the 'new' ROM will have to bear the following points in mind:

The addresses of the routines between 0000 and 0EE9 are unchanged.
Between 0F20 and 1022 they are moved up by 3 bytes.
Between 1046 and 1716 they are moved up by 4 bytes.
Between 1737 and 1DE1 they are moved up by 1 byte.
The character generator remains at 1E00.

The changes mean that in:

Chapter 4: the references to change are
- the CLEAR command routine is at 149A
- the 'assignment of a string variable' routine is at 13C8
- the FAST command routine is at 0F23
- the SLOW command routine is at 0F2B.

Chapter 6: the references to change are
- Floating point handling routine - 158A
- The function table - 1915
- The floating point calculator - 199D
- DIM - 1409   CLEAR - 149A   PAUSE - 0F32
- SLOW - 0F2B   FAST - 0F23

Chapter 7: the change to make is

As SLOW is now at 0F2B the value to be entered in location 16534 is changed to 43 decimal. Remember the checksum will go up by 3.

CONTENTS

Preface

In the spring of 1980 Science of Cambridge (now Sinclair Research) launched the ZX-80. It was the first of a new type of microcomputer for the hobbyist. At last there was a very cheap and reliable machine that together with an ordinary T.V. set and a cassette player formed a powerful home microcomputer system.

Although the ZX-80 was highly successful it was possible for Sinclair Research to put a much improved version on the market by the spring of 1981. This second machine is called the ZX-81.

The ZX-81 is supplied with a 8K ROM that contains the operating system program and a floating-point BASIC interpreter. The BASIC is very easy to use and is quite fast enough for most simple programming tasks. The added feature of 'syntax checking' is really quite a remarkable extra to be found on such a small machine.

However once BASIC has been mastered the attraction of machine code programming offers to the programmer, the possibility of producing programs that RUN at great speed, and can be as complicated, for their size, as any programs written for larger machines.

The main themes of this book are to develop an understanding of the Z80 machine code language and to discuss the actual workings of the 8K monitor program.

It is hoped that readers with only a knowledge of BASIC, will gain the ability to write short machine code programs for themselves and thereby derive even greater pleasure from their ZX-81.

1. Introduction

1.1 This book can be divided into two major parts.

The first part, chapters 1-4, discuss the Z80 microprocessor and its machine code instructions.

The second part, chapters 5-7, deals with actual machine code programs. Chapter 5 containing programs that illustrate the different types of machine code instructions in the Z80 instruction set. Chapter 6 discusses the 8K monitor program and chapter 7 gives some suggestions on how to go about writing a machine code program.

Throughout the book there are many references to the 8K monitor program. This program being chosen as an example program because it is the only machine code program that is supplied with the standard ZX-81.

It is assumed that the readers of this book will already have a fairly good understanding of the BASIC language as used in the ZX-81. New terms will be explained as they occur, drawing only on that knowledge of the BASIC language.

1.2 The standard ZX-81 microcomputer system:

The standard system comprises:

1. The ZX-81 mainboard with 1-16K RAM (or more)
2. The keyboard, (integral with the mainboard unless otherwise adapted).
3. The T.V. receiver.
4. The cassette player.

This standard system can be shown diagrammatically as;

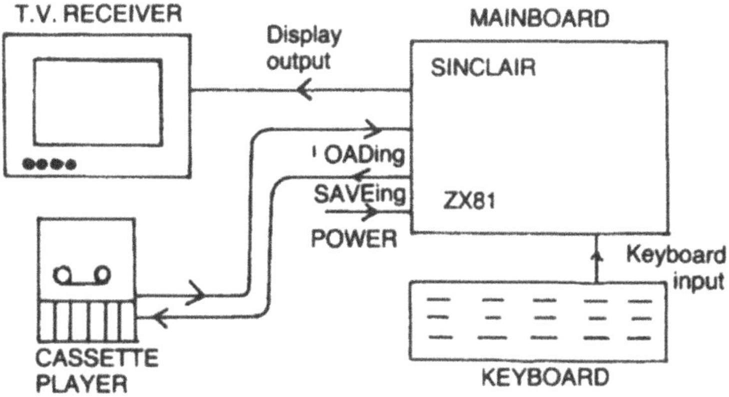

Diagram 1. The standard ZX-81 microcomputer system

Before the power supply is connected to the mainboard, it is really quite obvious that no work can be performed by the system. However once the power is connected the Z80 microprocessor starts working – executing instructions sequentially – at its operating speed of 3.25 mhz. A simple instruction will take 1.23 microseconds and the longest instructions no longer than 7.46 microseconds. The microprocessor then continues to execute between **141,00** and **812,500** machine code instructions each and every second until the power is turned off.

This book aims to develop an understanding for the reader of just what the computer system is doing all the time, and why even very simple tasks sometimes take a lot longer than 'one eight hundred thousandth' of a second to be completed.

This book also introduces the reader to the subject of monitor programs but let it suffice for the moment to say that it is the monitor program that is followed by the Z80, unless the BASIC USR command is being used, and it directs the ZX-81 to;

"Produce" a screen display.

"Scan" the keyboard to detect keystrokes.

"Provide" a system for the LOADing and SAVEing of programs on cassette tape.

"Give" the user the BASIC language.

2. The Z80 Microprocessor

2.1 The Z80 in outline.

The ZX-81 microcomputer has as its largest and most important 'silicon' chip a Z80 microprocessor. The Z80 is so called because it was developed by Zilog, Inc. of California, U.S.A. This company expanded and improved an earlier microprocessor, the INTEL 8080. The figure '8' in the name also implies that it is an eight bit microprocessor.

A microprocessor is a 'silicon' chip, and in common with other 'chips', it has input lines (= wires) that carry electrical impulses into the chip, output lines that carry impulses away, and power and ground connections.

But a microprocesor is a very specialised chip, that as the name implies has been designed to perform specifically as a small 'processor' or computer.

Internally it is amazingly complicated, but fortunately the internal structure can be divided into five functional parts. These are the Control Unit, the Instruction Register, the Program Counter, the 24 User-registers and the Arithmetic-logic unit.

This simplified view of the internal structure of the Z80 is shown diagrammatically in diagram 2.

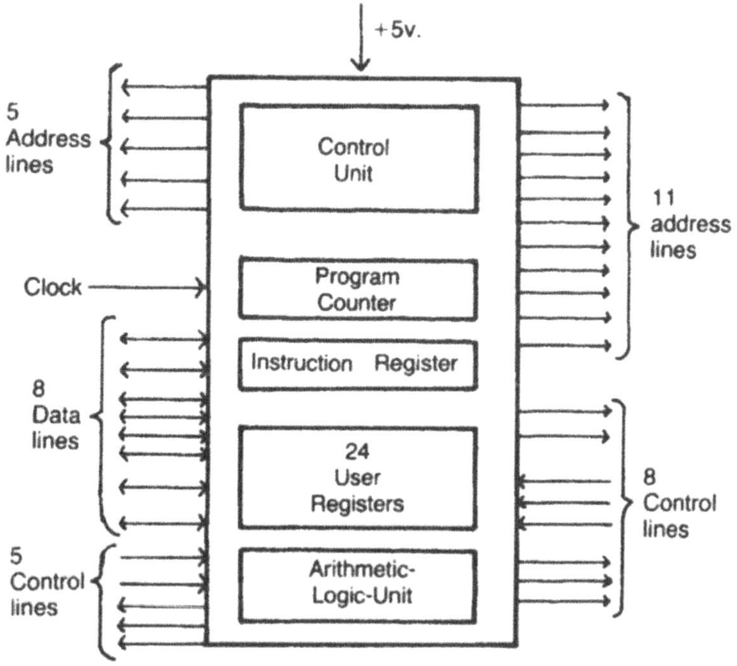

Diagram 2. The Z80 microprocessor in outline.

A Z80 microprocessor only works as a computer because it is a machine capable of following a 'stored program'. This program is required to be in the form of a sequence of machine code language instructions, that the Z80 can execute, together with any data that is required. The whole of this program must be present in memory that can be accessed by the Z80.

The 8K monitor program is such a program, and by being supplied in a ROM (read only memory) it is permanently present in memory. Indeed the standard ZX-81 has the monitor program so placed in memory that when the power to the Z80 is turned on, it is this program that is followed immediately.

2.2 The Data and Address buses

The Z80 microprocessor cannot work in isolation and it must therefore have connections to the other parts of the system. These connections are of three types, viz. the Control lines, the Data lines and the Address lines.

The Control lines are single tracked (= 1 wire), whereas the Data lines are usually collected together as an 8 track DATA BUS and the Address lines as a 16 track ADDRESS BUS.

The Control lines will be discussed in the next section.

The Data bus is used to carry 8 binary digits at a time, in parallel. The level of the voltage on a particular line signifying whether the line is carrying a binary 0 or a binary 1. Each binary digit is usually called a 'bit' and a collection of 8 bits held together for a particular purpose is called a 'byte'. The Data bus is therefore said to be able to carry '1 byte' of data at a time. The term 'byte' is commonly used to describe a 'place' where 8 bits of data would fit, rather than to an actual 'byte' of data.

The Address bus has 16 tracks and is therefore described as being '2 bytes' in width.

It is a fundamental principle of the Z80 that data is held as a byte of 8 binary digits and an address is held **always** as 2 bytes, that is 16 binary digits.

The Data bus is used to carry bytes of data to and from the Z80 and it therefore physically links the Z80 to the RAM chips (random access memory) and the ROM. The Data bus is also joined to the circuitry of the keyboard, the T.V. display and the cassette player interfaces.

The Address bus also links the Z80 to the RAM chips and the ROM, and in the ZX81 there are certain links to the keys of the keyboard.

Diagram 3 shows the Z80 and the main bus system.

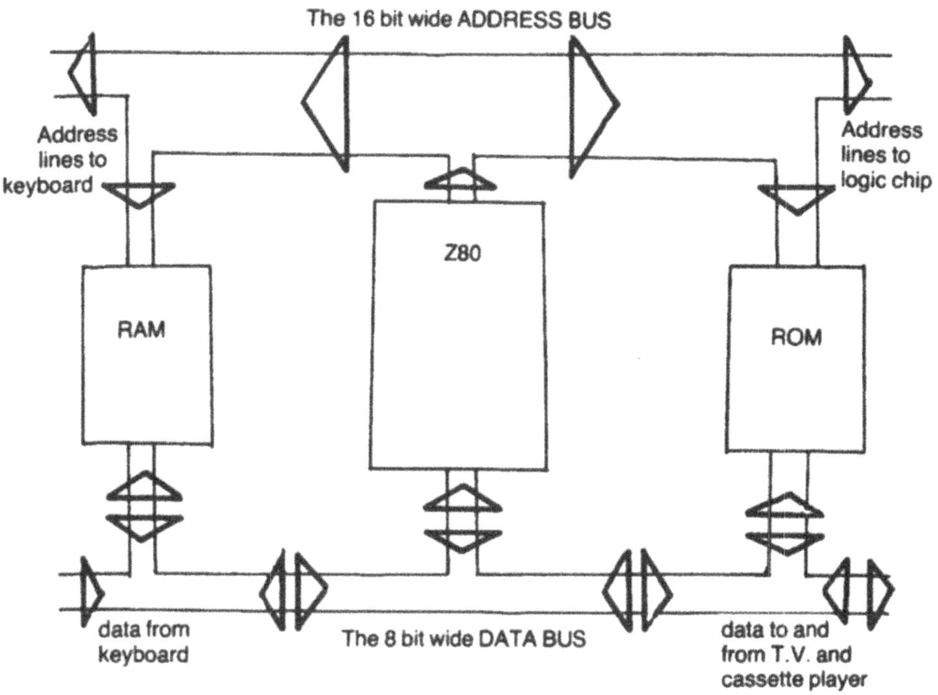

Diagram 3. The Address bus, the Data bus and the Z80.

The Address bus is used to hold the addresses of the location in the memory of the Z80, whether it be RAM, ROM or something else. As the Address bus can hold a 16 bit binary number it follows that the range of addresses that can be present on the Address bus can range from;

binary 0000 0000 0000 0000 – 1111 1111 1111 1111 which has the decimal equivalent of;

0 – 65535 (location 0 being the first location)

When the Z80 is running, bytes of data are continually being READ from memory and WRITTEN into the memory.

In order to perform a READ operation, the specific address of the required location in memory must first be placed on the Address bus, then the byte of data in that location can be copied and the copy passed along the Data bus to the Z80.

In order to WRITE a byte of data into the memory, the required address must first be placed on the Address bus, then the Z80 puts the byte of data onto the Data bus and the memory chip subsequently collects the byte of data.

The five functional parts of the internal structure of the Z80 will now be discussed in turn.

2.3 The Control Unit.

The Control unit of the Z80 can be likened, in a simplistic manner, to the 'manager of a production line'. It is therefore the responsibility of the Control unit, the manager, to arrange that materials (data) are brought into the Z80, that finished products (also data) are sent out to the correct destination and to ensure that the 'production' is timed successfully.

In the case of the Z80 there are a large number of different timing signals that are generated. Some of these signals are used only within the Z80 itself, with the rest being put out on the Control Lines.

The Control Lines are those lines that carry 'signals' to and from the Z80. For example there is a control line called 'READ' that is used during the operation of reading data from outside the Z80.

It is important to understand that the Control Unit, like the production manager, is in no way responsible for deciding which work is to be done, only for actually doing the work. The Z80 has to follow the 'program' as written by the programmer and the production manager has to follow the 'program' as set out by his company directors.

A more detailed discussion on the Control Unit and the Control signals is beyond the scope of this book.

2.4 The Instruction Register

The term 'Register' is used to describe a single byte location within the Z80 itself. It therefore is an actual place where 8 bits of data can be held. In the Z80 there are a whole series of registers and the moving of bytes of data 'into and out of' registers is the most important single feature of machine code programming.

The Instruction Register is a register within the Z80 that has the specific purpose of holding a copy of the instruction that is currently being executed.

As was said earlier, the Z80 microprocessor only works as a computer because it is a machine capable of following a stored program.

When the Z80 is following such a program, a copy of each instruction in turn will be placed in the Instruction Register prior to its being decoded and executed by the Z80.

2.5 The Program Counter

The Program Counter is not a single register but is a pair of registers within the Z80. It is used for the specific purposes of holding the address of the location in memory either of the current instruction that is being 'executed', or of the next instruction to be 'fetched'.

When an instruction is to be 'fetched', the Control Unit arranges that a copy of the contents of the location that is addressed by the Program Counter, is loaded into the Instruction Register. It is also one of the jobs of the Control Unit to ensure that the value in the Program Counter is then changed so as to 'point' to the location of the next instruction.

The actions of the Program Counter are very similar to those of the BASIC interpreter's 'Line number of current statement variable. (16391-2). This variable holds the line numbers of the 'current statements' as the interpreter goes through the lines of a BASIC program.

2.6 The User-Registers. (Main registers)

There are 24 User-Registers within the Z80. They have been termed 'User' registers because they can be filled with specified bytes of data by the programmer.

The names given to these 24 different registers are not at a first glance logically arranged. The reason for this state of affairs being that the Z80 is a microprocessor that has evolved from earlier, and less complicated, models. Certain of the names hark back to the earliest microprocessors that were ever made, whilst later names have been added in an 'ad hoc' fashion. Some names proving to be more appropriate, and informative, than others.

All of the registers, strictly, are single byte registers but they are commonly used as register pairs.

The following diagram shows the 24 User-registers of the Z80 displayed as 12 register pairs, The 'bit numbers' are also shown.

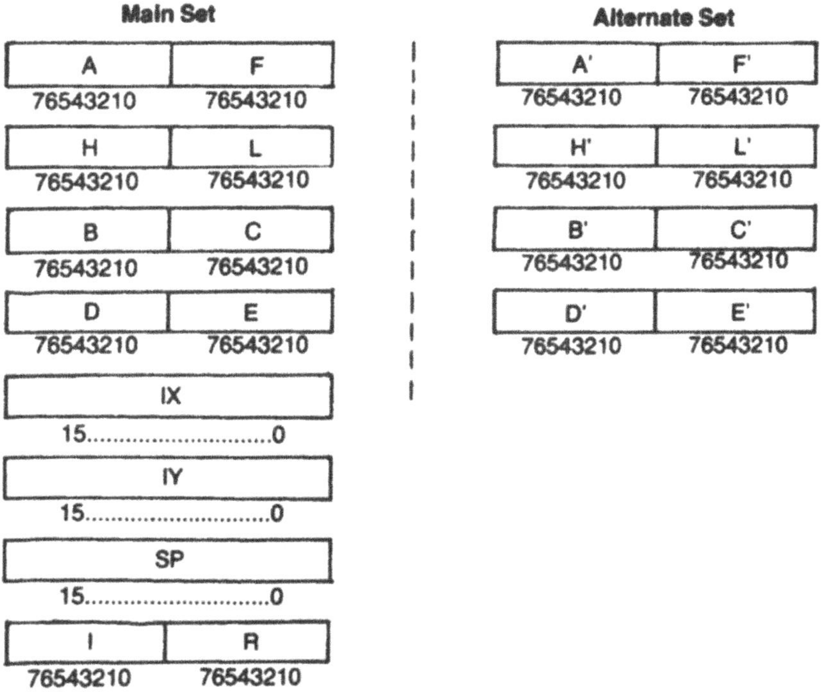

Diagram 4. The 24 User-Registers of the Z80.

Each of these registers will now be discussed briefly:

The A register:

This register is the single most important register of the Z80. It is often called the ACCUMULATOR, a name that goes back to those models in which there was only one register that could be used to 'accumulate' a result.

In the Z80 the A register is extensively used for arithmetic and logical operations, and indeed there are many operations that can only be performed using the A register.

There is a great number of different ways in which a byte of data can be entered into the A register by the programmer and hence there are many machine code language instructions that involve the use of the A register.

The F register:

This is the FLAG REGISTER and it is often considered to be a collection of 8 bits rather than a true register.

The concept of flags will be dealt with fully in chapter 4, but simply a flag is a 'bit' that is given the value 0 (RESET) or the value 1 (SET) depending on the result of the operation.

The flag register does have 8 bits but the programmer is really only concerned with the '4 major flags'. These are called the Zero flag, the Sign flag, the Carry flag and the Parity/overflow flag.

The 'minor flags' are used by the Control Unit and cannot directly be used by the programmer.

The HL register pair:

In early microprocessors the register that was used to hold addresses was a single byte 'address register', and was capable of addressing 256 locations. However when 2-byte address registers were introduced, one of the registers was called the 'high address register' and the other the 'low address register'. A register pair is capable of addressing 65536 locations.

The 'H' and 'L' therefore derive their origins from the words 'high' and 'low'. It is also interesting to note that the 'high' register, being a later development has lead to the situation whereby an address, is always given as the 'low' byte followed by the 'high' byte.

In the Z80 the HL register pair is just one of three register pairs that are used to hold addresses. However the HL register is the most important. The HL register pair can also be used to hold 16 bit numbers, rather than addresses, and there are a certain number of arithmetic operations that can be performed on these numbers. The registers of this pair can also be used as single byte registers. However there are relatively few operations that can be performed on the H register, or the L register, as compared to the A register.

The BC and DE register pairs:

These are the other register pairs that are predominantly used within the Z80 to hold addresses. It would appear that their names were simply derived because an 'A' register already existed.

Once again the programmer may use the registers as single registers, and also there are certain instructions that use some of these registers as counters.

The Alternate register set:

The Z80 is an interesting microprocessor in that it has an alternate register set for the A, F, H, L, B, C, D & E registers. These alterate registers are designated for A', F', H', L', B', C', D' & E' registers.

There are two special instructions in the Z80 instruction set that allow for the contents of the alternate set of registers to be exchanged with the contents of the current main set of registers. Once the registers have been exchanged the Z80 will work with the 'former alternate set' believing it to now be its 'main set'. The 'former main set' will now be treated as an 'alternate set'.

The programmer may exchange the register sets, totally or in part, as often as he wishes in a particular program.

The concept of there being alternate registers may sound to be .ery simple, but in practice it is far from being so. The problem is that the programmer has to make sure that he knows which set of registers is actually being used at a particular moment, as there are no machine code instructions that only work on one set of registers and not the other.

Consequently there are many programmers that never use any of registers in the alternate set, and their programs are perfectly successful. However programs that take full advantage of the alternate set of registers may run faster than programs that do not.

In many Z80 systems the use of the alternate set of registers is restricted, and indeed this is the case in the ZX-81 system running with the 8K monitor program. This means that a machine code program that is to 'return' to BASIC must not use some of the alternate registers, however a self-contained machine code program is quite free to use the full set of alternate registers.

The IX and IY register pairs:

These two register pairs are used to perform operations using 'indexing'. This is a facility which allows for entries in a list or table to be manipulated as long as the Index Register pair being used holds the 'base' address of the table or list, and the position of the required entry is known relative to the 'base' address.

The Stack Pointer:

The Stack Pointer is a register pair that is used to 'point' to a location in memory in an area called the 'stack'.

All microprocessors in current use require a stack, that is an area of memory to use as a working space for the storage of addresses or data on a temporary basis. Such a working area is called a stack because each entry is stacked next to the previous one.

The Z80 uses a stack that 'grows downwards' in memory, so an analogy might be to a high rise block of appartments in which the first tenant moves into the top appartment, the next tenant into the one below and so on downwards. The stack is used on a first-in last-out principle, so the first tenant to move out will always be the latest tenant to have moved in.

The stack pointer is used to point to the different locations in the stack in a very special way. The stack pointer always holds the address of the last location to have been filled. Therefore when a new entry is to be made, the Control Unit first arranges for the value of the stack pointer to be decreased by one (decremented) so as to point to the actual location that is to be filled. The required data is then moved to this location.

When data is being removed from a stack the stack pointer has to be increased in value.

To be a little more precise the stack pointer is always decremented twice when data is added to the stack, and incremented twice after data has been removed from the stack. This is because all data movements involving the stack require the handling of two separate bytes of data.

It is an important point that strictly data is not removed from the stack but only a copy of the data is made, and the data still remains in the locations until it is overwritten at a later date.

To the programmer the use of the stack is almost a challenge. Once again there are many programmers that shy away from using the stack to temporarily hold data, preferring to use ordinary memory locations instead. Programs that do not use the facility of the stack may not run quite as fast as those programs that do use the stack because moving data to and from the stack is very quick.

The I register:

This is the Interrupt Vector register. The Z80 in its normal running state executes sequentially the instructions in a machine code program. However this 'normal' execution of a program can be 'interrupted'. In most Z80 systems the 'interrupts' would be generated at the request of such devices as printers, disk units and clocks. However in the ZX-81 system 'interrupts' are only used when forming the T.V. picture.

When a device wishes to interrupt the Z80 it places a signal on the appropriate control line. The Z80 then responds by stopping its execution of the normal program and attends to the request of the device.

In the ZX-81 system an 'interrupt' will lead to the execution of either the routine that is located at address 0038, or the routine that is located at address 0066.

However in larger systems many interrupt handling routines may be needed and the Interrupt Vector Register is used to hold the 'high byte' of the address of a 'table of addresses' for the different handling routines. When the I register is used in this way up to 128 different addresses can be held in a single table.

The R Register:

This is the Memory Refresh Register. This register is just a simple counter that is incremented every time an instruction or a byte of data is fetched from the memory. The value held in the register therefore alters between 0 and 255, over and over again.

In most Z80 systems this register is used to indicate which locations in memory need to be refreshed (recharged) at a particular time. However in the ZX-81 system the R register is used to count the number of characters that go to form a single line of the T.V. display.

2.5 The Arithmetic-Logic Unit:

The Arithmetic-logic Unit, the ALU, is the last of the functional blocks of the Z80, and it is concerned, as its name implies, which arithmetic and logical operations.

It is important to realise that the actual operations that can be performed by the ALU are very limited in scope.

Simply binary addition and subtraction are possible, but not binary multiplication or division, as these latter operations are very complex in binary arithmetic. Incrementation (adding 1) and decrementation (subtracting 1) are just special cases of addition and subtraction and are readily managed. The unit is also able to perform a large number of 'bit' operations, by this is meant that it can assess the value of a single bit from within the byte held in the unit at a particular moment.

All of the above operations are discussed fully in chapter 4.

2.6 The Concept of machine code instructions:

Now that the 'hardware' has been discussed it is appropriate to discuss the actual structure of a machine code program. Machine code program instructions are usually written in an 'assembler language' format, this means that 'mnemonics' (words) are given to each instruction so as to make it easier to read. Otherwise a machine code program would appear as a simple list of numbers in binary, hexadecimal or decimal arithmetic.

There is fortunately a great similarity between the structure of a BASIC program and that of a machine code program.

A BASIC program is made up of a set of BASIC lines, each with a line number that shows its correct position in the program, and in turn each BASIC line is made up of an initial command followed, if necessary, by further data.

A machine code program can have its structure described in exactly the same way. A program is made up of a set of instruction lines, each of which has allocated to it, an address in memory where the line is to be placed, and in turn the instruction line is made up of an initial instruction code followed, if necessary, by further data.

In the ZX-81 BASIC there are only 31 commands but in the Z80 machine code langauge there are over 500 different instructions. Fortunately though, the situation is helped by the instructions being easily divided into 18 smaller groups.

The Z80 instruction codes are 1-2 bytes in length, and of the 256 different numbers that can be held in a single byte, 252 numbers correspond to 1-BYTE INSTRUCTIONS. The remaining 4 numbers are used to form 2-BYTE INSTRUCTIONS.

Throughout this book the term 'instruction code' is reserved for describing the 1, or 2, bytes of the actual instruction, and where data has to be placed after the instruction code, this data will always be described as 'data'.

The full instruction length of the instructions in the Z80 machine code langauge is said to be between 1 and 4 bytes, as there are no instructions that taken together with any required data occupy more than 4 bytes.

Each of the different instructions has its own mnemonic, and these have been chosen to explain what the instruction is actually doing. E.G. a simple instruction mnemonic is 'LD A, B' which can be expanded to give load a copy of the contents of the B register into the A register'.

To illustrate the points made in this section the following diagram shows a few lines of machine code, together with the 'assembler' mnemonics and an attempt using BASIC to show just what the lines achieve.

Indeed practically all machine code programs can be described in a PSEUDO-BASIC, and to the programmer new to machine code language programming, the approach can be very helpful.

Machine Code Language			BASIC Language
address	Hex.code	mnemonic	line No. Line
052B	3E 78	LD A,+78	1323 LET A=120
052D	5F	LD E,A	1325 LET E=A
052E	21 82 04	LD HL,+0482	1326 LET HL=1154
0531	19	ADD HL,DE	1329 LET HL=HL+DE
0532	19	ADD HL,DE	1332 LET HL=HL+DE
0533 .... continues			1333 ..... continues

Diagram 5. An Introductory Machine Code Program

3. The Simple Mathematics

Absolute binary arithmetic

As it has already been said, the Z80 holds numbers which it is manipulating, either as 1 byte of 8 binary digits, or less commonly, as 2 bytes of 16 binary digits.

By saying that a number is held as an absolute binary number it is meant that each byte has the binary range of 0000 0000 – 1111 1111, or the decimal range of 0-255, and importantly there is no way that such a number could ever be considered to be negative.

It is important to realise that all numbers held in the Z80 are **always** absolute binary numbers. To a certain extent this can be shown by using the BASIC PEEK command, as this command will return the contents of any of the memory locations as a decimal number in the range 0-255.

The following program shows this in a very simple way:

```
10 SCROLL
20 PRINT PEEK (RND*65535)
30 GOTO 10
RUN
```

The program prints the PEEK values and it can be seen that they all lie in the range 0-255 decimal.

It is also important to realise that when operations, such as addition, take the contents of the byte past 255 then the contents will revert to 0, instead of going past 255. An operation such as subtraction will cause the contents of a byte to return to 255 every time 0 is reached.

This behaviour of an absolute binary number can be shown by the following BASIC programs:

For the addition of '255+5'

```
10 LET A=255
20 LET A=A+255
30 IF A>=256 THEN LET A=A-256
40 PRINT A
```

For the subtraction of '5−8'

```
10 LET A=5
20 LET A=A-8
30 IF A<0 THEN LET A=A+256
40 PRINT A
```

In a Z80 system all of the numbers are in 'absolute binary', but often the programmer wishes to place a different interpretation on the value of the number. The commonest method in use at the present time is called '2's complement arithmetic', and indeed in the Z80 instruction set there are several instructions that have been included to help the programmer handle this form of arithmetic.

3.2 2's Complement Arithmetic

The concept behind '2's complement arithmetic' is very simple, but when it is actually used in a program the results can be very confusing. All machine code programmers must therefore try to become familiar with this important form of arithmetic.

Simply the method enables the programmer to designate the numbers in the binary range 0000 0000 – 0111 1111 to the decimal range of 0 to +127, and the numbers in the binary range 1000 0000 – 1111 1111 to the decimal range −128 to −1.

A result of this interpretation is to make 'bit 7' (the left-hand most bit) act as a 'sign' bit. This bit will have the value 0 (Reset) for positive numbers and the value 1 (Set) for the negative numbers.

The following diagram illustrates the method.

	BINARY	DECIMAL	HEX.
Positive numbers	0111 1111	+127	7F
	0111 1110	+126	7E
	⋮	⋮	⋮
	0000 0010	+2	02
	0000 0001	+1	01
	0000 0000	0	00
Negative Numbers	1111 1111	−1	FF
	1111 1110	−2	FE
	⋮	⋮	⋮
	1000 0001	−127	81
	1000 0000	−128	80

↑ Sign bit

Diagram 6. 2's complement arithmetic (1 byte)

Fortunately there is an easy way to find the 2's complement value of the decimal numbers −128 to −1, and that is as follows:

a) form the binary number of the absolute decimal number. e.g. to find the 2's complement of −45, first find the binary equivalent to +45, which is 0010 1101.

b) form the 1's complement of this number, (simply change all the bits to their opposite values) e.g. 0010 1101 is changed to 1101 0010.

c) add 1, (as you have passed zero) e.g. 1101 0010 + 0000 0001 = 1101 0011

The reverse of the above method can be used to convert 2's complement negative numbers to their decimal equivalents.

3.3 Hexadecimal coding

All machine code programmers do find that any attempt to produce more than just a very few bytes of binary coding is just not practical. Therefore programmers usually use a form of shorthand for describing binary numbers and the commonest system to use is 'Hexadecimal coding'. (Hex. coding).

In many microcomputer systems the resultant 'Hex-code' can be entered via the monitor program directly into the computer, however the 8K monitor program of the ZX-81 does not have this facility so the programmer has either to enter his machine code in decimal numbers, or to write for himself a Hex-loader routine in BASIC.

The principle behind Hex. coding is once again very simple, but it takes a very long time to become fluent in its use, and even programmers of some years experience still have trouble.

To obtain the Hex. code for a 8 bit binary number, the number is split into 2 groups of 4 bits, each group being called a 'nibble'. A Hex. character is then assigned to represent each 'nibble'. Hex. characters are the decimal numbers 0-9 and the letters A-F.

The following table shows the different Hex. characters;

Binary	**Decimal**	**Hex. character**
0000	0	0
0001	1	1
0010	2	2
0011	3	3
0100	4	4
0101	5	5
0110	6	6
0111	7	7
1000	8	8
1001	9	9
1010	10	A
1011	11	B
1100	12	C
1101	13	D
1110	14	E
1111	15	F

A single binary byte is therefore represented by a pair of Hex. characters.

e.g.	**binary**	**Hex. code**
	1010 1111 =	AF
	and	
	0001 1110 =	IE

A 2-byte binary number will be represented by 4 Hex. characters.

e.g.	**binary**	**Hex. Code**
	1000 1000 0000 1011 =	880B
	and	
	0001 0000 1111 0101 =	10F5

In the rest of the book 2-byte addresses will be shown as a single group of 4 Hex. characters.

The following examples show how a 4 Hex. character number can be converted to a decimal number.

Example Hex. to decimal

Hex. number = 789A

Decimal equivalent =

```
  7 * 4096 = 28672
  8 *  256 =  2048
  9 *   16 =   144
+ A *    1 =    10
------------------
789A       = 30874
------------------
```

or if the Hex. characters are taken in pairs:

```
Hex. 78 = 120 ; 120 * 256 = 30720
Hex. 9A = 154 ; 154 *   1 =   154
                          -------
                          = 30874
```

Appendix iii. is a Decimal-Hex. conversion table and the use of such a table may be quicker and easier than trying to convert numbers by the above method.

4. The Z80 Machine Code Instruction Set

4.1 Instructions and data

The stage has now been reached when the actual instructions of the Z80 machine code instruction set can be discussed in turn.

The instructions are divided, in this book, into 18 groups, with each group containing the instructions that hold a strong resemblance to each other.

This book follows the convention that distinguishes between the actual instruction, the initial 1-2 bytes, of the instruction line and the data, up to 2 bytes, that if required is placed after the instruction proper. The total number of bytes in an instruction line being therefore 1, 2, 3 or 4.

There are six classes of data that may follow instructions. They are:

1. A single byte constant. (+dd)
   i.e. a number in the range Hex. 00-FF, decimal 0-255.
   Instructions that are requred to be followed by a single byte constant have this indicated in their mnemonics by there being a '+dd'.
   e.g. LD A,+dd
   This instruction loads the A register with a constant held in the next byte.
2. A 2-byte constant. (+dddd)
   i.e. a number in the range Hex. 0000-FFFF, decimal 0-65535.
   Instructions that are required to be followed by a 2-byte constant have this indicated in their mnemonics by there being a '+dddd'.
   e.g. LD HL,+dddd
   This instruction loads the HL register pair with the constants held in the next 2 bytes. The first byte going into **L**, and the second going into **H**.
3. A 2-byte address. (addr.)
   i.e. a number in the range hex. 0000-FFFF, decimal 0-65535.
   Instructions that are required to be followed by 2 bytes of data that will be used as an address have this indicated in their mnemonics by there being a 'addr.'.
   e.g. JP addr.
   This instruction causes an absolute jump to the 'address' held in the next 2-bytes. (=GOTO) The first byte holds the **low** byte of the address and the second byte the **high** byte of the address.

4. A single byte displacement constant. (e)
   i.e. a number in the range Hex. 00-FF, decimal −128 to +127, as the number is always considered to be in 2's complement arithmetic.
   Instructions that are required to be followed by a single byte displacement constant have this indicated in their mnemonics by there being a 'e'.
   e.g. JR e
   This instruction causes a 'relative jump', the displacement constant indicating the size of the jump.
5. A single byte indexing displacement constant. (d)
   i.e. a number in the range Hex. 00-FF, decimal −128 to +127, as the number is always considered to be in 2's complement arithmetic.
   Instructions that are required to be followed by a single byte indexing displacement constant have this indicated in their mnemonic by there being a 'd'.
   e.g. LD A,(IX+d)
   This instruction loads the A register with the contents of the location whose address is formed by the addition of 'd' to the current value held in the index register pair, IX.
6. A single byte indexing displacement constant AND a single byte constant. (d,+dd)
   i.e. Two bytes of data each in the range Hex. 00-FF, decimal −128 to +127 for the first byte and 0-255 for the second byte.
   Instructions that are required to be followed by two bytes of data for this purpose have this indicated in their mnemonics by there being a 'd' and a '+dd'.
   e.g. LD (IX+d),+dd
   This instruction loads the constant '+dd' into the location, whose address is formed by the addition of 'd' to the current value held in the index register pair, IX.

4.2 The Instruction Groups

There are many ways in which the hundreds of different machine code instructions could be split into groups. The method chosen in this book is to split the instruction into functional groups, so that the reader can study the instructions in a group and then RUN the BASIC programs from chapter 5 that demonstrate those instructions.

Group 1. The No operation and Return instructions:

mnemonic	**instruction** Hex.
NOP	00
RET	C9

It may initially be strange to find that these two instructions should form the first group. However these instructions can be used alone to make **complete** machine code programs that will RUN under the BASIC USR command.

The NO OPERATION instruction when executed simply results in the Z80 marking time for 1.23 microseconds. It is used just as 'padding' in a program or to cause short timing delays when they are needed.

The RETURN instruction has exactly the same effect as the BASIC RETURN command, as it used to make a return from a subroutine. If the machine code program is being RUN under the USR command then the final RETURN instruction will cause a return to BASIC.

The actual steps of the execution of the RETURN instruction are to load the Program Counter register pair with two bytes of data taken from the stack. The Stack Pointer register pair will have its contents incremented twice because two bytes of data have been removed. The first byte taken off the stack forms the low address byte and the second byte forms the high address byte of the Program Counter.

A more detailed discussion of the RETURN instruction is to be found on page 65.

The BASIC programs that demonstrate these instruction are to be found on page 83.

Group 2. The instructions for loading registers with constants:

The following instructions are all involved with loading the registers with single byte constants:

mnemonic	**instruction** Hex.
LD A,+dd	3E dd
LD H,+dd	26 dd
LD L,+dd	2E dd
LD B,+dd	O6 dd
LD C,+dd	OE dd
LD D,+dd	16 dd
LD E,+dd	1E dd

The instruction length of all these instructions is 2 bytes, one byte for the instruction proper and one byte for the constant.

The effect of all the above instructions is simply to load the specified register with a copy of the constant.

The following instructions are all involved with loading register pairs with 2-byte constants:

mnemonic	**instruction** Hex
LD HL,+dddd	21 dddd
LD BC,+dddd	01 dddd
LD DE,+dddd	11 dddd
LD IX,+dddd	DD 21 dddd
LD IY,+dddd	FD 21 dddd
LD SP,+dddd	31 dddd

The instruction length of these instructions will be either 3, or 4 bytes.

The effect of the above instructions is to load the specified register pair with a copy of the data held in the 2 bytes held after the instruction proper. The first byte going to the low register, (i.e. L, C, E, X, Y or P) and the second byte going to the high register. (i.e. H, B, D, I or S).

The instructions in this 'Group 2' are some of the most commonly used instructions in any program.

The loading of single byte constants is used for many different reasons. The following example, taken from the 8K monitor program shows just one use of this type of instruction.

An example from the 8K monitor program

In the 'print a whole BASIC line routine' it is necessary when printing the 'cursor' to determine whether it should be a 'F', 'G', 'K' or a 'L' (inverted). Instructions that load the B register with the character codes for 'inverse F' or 'inverse K' are used. When 'inverse G' or 'inverse L' are required, they are obtained by changing the contents of the B register.

address	Hex. code	mnemonic	comment
O7AO	O6 AB	LD B,+AB	the 'inverse F'
		and	
O7A8	O6 BO	LD B,+BO	the 'inverse K'

The instructions in 'Group 2' that load register pairs with 2-byte constants are also very important. The constants can either represent numbers in the decimal range O-65535, or represent addresses for any location in memory.

The HL, BC and DE register pairs are the 'working' pairs, whereas the IX, IY and SP register pairs are used to perform 'special' functions.

The loading of constants into the 'working' register pairs is very straightforward, but the use of the indexing register pairs is worth discussing further.

The IX and IY index register pairs are used to 'index' along a list, a table or just a block of code. The actual usage of the IY register pair in the 8K monitor program forms a good illustration of how the IY register pair can be used.

An example from the 8K monitor program.

In the monitor program, at location Hex. 03F8 is an instruction line that results in the IY register pair being set to Hex. 4000, decimal 16384.

address	Hex. code	mnemonic	comment
O3F8	FD 21 OO 4O	LD IY,+4OOO	Dec. 16384

The purpose of this line is to set the IY register pair to hold the 'base address of the system variables'. The value of the contents of the IY register pair thereafter remains unchanged, and the IY register pair is frequently used to 'index' through the system variables. e.g. the location 16385, Hex. 4001, that holds various flags, is referred to as 'IY+1' where the value '1' is the displacement constant.

The BASIC programs that demonstrate these instructions are to be found on page 84.

Group 3. Register copying and exchanging instructions:

There really is a most comprehensive set of instructions for copying the contents of one register into another. There are also three instructions for the copying of the contents of register pairs into the Stack Pointer.

The following table gives the instruction codes for all the instructions that copy the contents of a single register 'r' into another specified register.

r register	LD A,r	LD H,r	LD L,r	LD B,r	LD C,r	LD D,r	LD E,r
A	7F	67	6F	47	4F	57	5F
H	7C	64	6C	44	4C	54	5C
L	7D	65	6D	45	4D	55	5D
B	78	6Ø	68	4Ø	48	5Ø	58
C	79	61	69	41	49	51	59
D	7A	62	6A	42	4A	52	5A
E	7B	63	6B	43	4B	53	5B

LD	A,I	ED	57
LD	A,R	ED	5F
LD	I,A	ED	47
LD	R,A	ED	4F

Table of the 53 single register-to-register copying instructions.

All the instructions in the main part of the table are very commonly used instructions. They are all single byte instructions and because there is no requirement for the Z80 to fetch any data from outside the microprocessor itself, the instructions are all executed very quickly. (1.23 microseconds).

The special instructions for handling the contents of the Interrupt Vector Register and the Refresh Register are not used by the programmer in most Z80 systems.

However, both of these specialised registers are used in the 8K monitor program as detailed next.

An example from the 8K monitor program.

In the ZX-81 system the Interrupt Vector Register is used to form the high part of the address of the characters in the 'character generator'. The 'character generator' is the part of the 8K ROM that holds the details of the shape of each of the characters that can be displayed on the T.V. screen. The base address of the 'character generator' is Hex. 1E00, and therefore the I register is loaded with Hex. 1E. The actual instruction lines are:

address	Hex. code	mnemonic	comment
03F2	3E 1E	LD A,+1E	Load A with a constant.
03F4	ED 47	LD I,A	Transfer it to I.

Although the 'LD I,A' instruction is a rather specialised instruction, it still is a typical 'single register-to-register copying instruction'.

The Refresh Register is also used in the 8K monitor program. It is loaded with a copy of the A register in the instruction lines:

address	Hex. code	mnemonic	comment
0041 & 02B5	ED 4F	LD R,A	Load R with a copy of A

The Refresh Register is then used as a counter for the characters of a line as they are passed to the T.V. When the register has counted 32 characters a hardware interrupt is generated.

The three instructions for copying the contents of a register pair into the Stack Pointer are:

mnemonic	**instruction** Hex.
LD SP,HL	F9
LD SP,IX	DD F9
LD SP,IY	FD F9

The use of these three instructions is specialised but the 'LD SP,HL' instruction is used in the 'initialisation routine' in the 8K monitor program.

An example from the 8K monitor program.

In the instruction line held at 03EC is the following code:

address	Hex. code	mnemonic	comment
03EC	F9	LD SP,HL	Set Stack Pointer.

The effect of this line is to set the Stack Pointer so as it points to an area of the RAM where the stack can be put.

In this 'Group 3' there are also three instructions that allow for the exchanging of the contents between registers. The instructions are:

mnemonic	**instruction** Hex.
EX DE,HL	EB
EXX	D9
EX AF,A'F'	O8

Of the three instructions only the 'EX DE,HL' instruction is concerned with the exchanging of the contents of registers solely within the main set of registers.

The 'EX DE,HL' is a very useful instruction as it allows for the contents of the two register pairs to be switched over. This is important because there are certain operations that can be performed on one register pair and not the other. One example of such a task is the addition of two 16 bit numbers that can only be performed using the HL register pair. Therefore if a number presently held in the DE register pair is to form part of a 16 bit addition,then the register pairs are exchanged, the addition performed and the registers exchanged back. This technique of 'exchanging-performing a task-exchanging again' is a very commonly used technique.

An example from the 8K monitor program.

In the 'Change all pointers' routine' it is, in effect, necessary to add the contents of the BC register pair to the contents of the DE register pair. Unfortunately there is no instruction to perform this operation using the DE register pair, but it can be done using the HL register pair. Hence the following method is used:

address	Hex. code	mnemonic	comment
O9CO	EB	EX DE,HL	Exchange.
O9C1	....	....	The addition.
O9C2	EB	EX DE,HL	Exchange again

which has the effect of: LET DE=DE+BC.

The other two instructions are involved with the handling of the registers in the alternate register set.

The 'EXX' instruction causes a switching over of the H,L,B,D and E registers with the H',L',B',C',D' and E' registers.

The 'EX AF,A'F'' instruction is simpler in that only the A and F registers are switched with the A' and F' registers.

Using of the alternate registers is always 'complicated' programming and the 8K monitor program uses the alternate registers frequently!

An example from the 8K monitor program.

The A' and F' registers are used to hold certain values involved in the production of the T.V. display in the 'slow' mode.

The following lines from the 'slow display routine' show the registers being exchanged, output signals being generated and the registers being exchanged back.

address	Hex. code	mnemonic	comment
0211	08	EX AF,A'F'	Exchange registers.
212 to 219			Output timing signals.
21A	08	EX AF,A'F'	Exchange registers back.

The other registers of the alternate register set are used frequently for two specific reasons. When printing a character the current values held in the registers are 'preserved' by exchanging register sets (but not AF) and again in the Calculator routines the alternate registers are used to 'preserve' values whilst other work is being done.

The BASIC programs that demonstrate these instructions are to be found on page 85.

Group 4. Instructions for the loading of registers with data copied from a memory location.

The Z80 instruction set has many instructions that 'fetch' data from the memory and then 'load' that data into a main register.

The address of where the data is to be found must be specified, as must the name of the register, or register pair, into which the data is to be placed.

The instructions in this group can be divided into three subgroups that depend on just how the addressing of the memory is performed.

There is:

(a) 'absolute addressing' when the actual address is held as data after the instruction proper.

(b) 'indirect addressing' when the required address is already held in the HL, BC or DE register pairs.

(c) 'indexed addressing' when the address is formed by the addition of a displacement value to a 'base' address already held in an index register pair.

SUBGROUP A: INSTRUCTIONS USING 'ABSOLUTE ADDRESSING'

The following instructions form this subgroup:

mnemonic	**instruction** Hex.
LD A,(addr.)	3A addr.
LD HL,(addr.)	2A addr. (usual form)
	or
	ED 6B addr.
	(unusual form)
LD BC,(addr.)	ED 4B addr.
LD DE,(addr.)	ED 5B addr.
LD IX,(addr.)	DD 2A addr.
LD IY,(addr.)	FD 2A addr.
LD SP,(addr.)	ED 7B addr.

The 'LD A,(addr.)' instruction is the only instruction in the whole of the Z80 instruction set that allows for a single byte of data to be copied into a single register. This instruction 'load the A register with a byte of data copied from a memory location' is really one of the oldest microprocessor instructions. It existed on the very earliest models. However in the Z80 it is just one of many instructions that is possible to use.

An example from the 8K monitor program.

As part of the 'initialisation routine' it is necessary to assess the size of the memory available, so as to determine whether the RAM is large enough to hold a 'complete display file'.

The instruction line held at O4B7 uses the 'LD A,(addr.)' to collect the value of the high byte of RAMTOP.

This value is then compared to the constant Hex.4D. The result of this comparison shows whether or not there is 3¼K available.

address	Hex. code	mnemonic	comment
O4B7	3A O5 4O	LD A,(4005)	High byte of RAMTOP.
4BA	....	....	Test the result.

Note carefully how the **low** byte of the address has to be placed before the **high** byte of the address.

The other instructions in this subgroup are very important instructions as they enable **two** bytes of data to be copied from memory into a register pair.

The mnemonics for these instructions really are abbreviated, it can be helpful to consider the mnemonics should be of the form;

e.g. LD BC, (addr). expands to LD C,(addr.) and LD B,(addr. + 1).

The expansion shows that the addressed byte is copied to the low register of the register pair, and that the **following** byte is copied into the **high** register of the register pair.

These instructions are very commonly used in any program and the following example shows just two of the hundreds of occurences.

An example from the 8K monitor program.

In the 'initialisation routine' a test is made to see if the 'current line with the program cursor' comes before the 'top program line in the listing'.

In order to do this test the system variables involved are loaded into the HL and DE register pairs.

address	Hex. code	mnemonic	comment
Ø41C	2A ØA 4Ø	LD HL,(4OOA)	'E-PPC'
41F	ED 5B 23 4Ø	LD DE,(4O23)	'S-TOP'

SUBGROUP B: INSTRUCTIONS USING 'INDIRECT ADDRESSING'

The instructions in this subgroup use the HL, BC or DE register pairs to point to the required memory location. There are instructions for loading the A register that use all these register pairs, but the other main registers can only be loaded using the HL register pair.

The instructions are:

mnemonic	**instruction** hex.
LD A,(HL)	7E
LD A,(BC)	ØA
LD A,(DE)	1A
LD H,(HL)	66
LD L,(HL)	6E
LD B,(HL)	46
LD C,(HL)	4E
LD D,(HL)	56
LD E,(HL)	5E

All of the instructions in this subgroup have an instruction length of 1 byte.

It is important to point out that these instructions result in the loading of just one byte of data into a single register.

These instructions are once again very common instructions and the example below shows instructions from this subgroup being used twice to load a register pair with two bytes of data that were held together in the memory.

An example from the 8K monitor program.

The 'Change all pointers routine' is a very frequently used routine. It is used to increase or decrease the values of the 9 main pointers. i.e. D-File, DF-CC . . . STKEND, which are the system variables 16396 to 16413.

The pointers require to be changed whenever characters are added, or removed, from the program area.

In order to pick up the pointers in turn, the HL register is set to the 'base address' of the list and then stepped along the list. 'LD E,(HL)' and 'LD D,(HL)' instructions are used to 'pick-up' the required values.

The following instruction lines show just the 'pick-up' part of the routine.

address	Hex. code	mnemonic	comment
Ø9AF	21 ØC 4Ø	LD HL,+4OOC	Set base pointer
9B2	3E Ø9	LD A,+Ø9	9 variables
9B4	5E	LD E,(HL)	Low byte to E
9B5	23	INC HL	Increase pointer
9B6	56	LD D,(HL)	High byte to D

(the 'INC HL' instruction will be explained in more detail on page 39.)

SUBGROUP C: INSTRUCTIONS USING 'INDEXED ADDRESSING'.

The instructions in this subgroup allow the programmer to load the main single registers with a copy of a byte of data specified as being held in a table, list or just a block of code. The base address already being held in the IX or IY register pairs.

The following diagram shows how the different values of the displacement constant are used to form the addresses of the entries in the table.

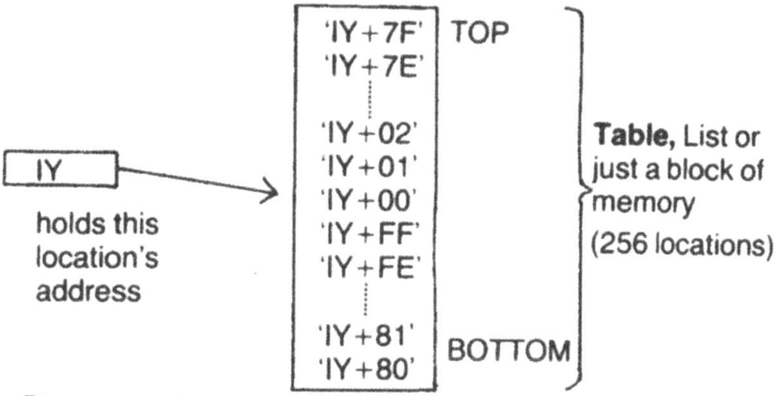

Diagram 7. To show the IY register pair being used to address a table of 256 locations.

The actual instructions in this subgroup are:

mnemonic	**instruction** hex.
LD A,(IX+d)	DD 73 d
LD H,(IX+d)	DD 66 d
LD L,(IX+d)	DD 6E d
LD B,(IX+d)	DD 46 d
LD C,(IX+d)	DD 4E d
LD D,(IX+d)	DD 56 d
LD E,(IX+d)	DD 5E d

Note for the IY Instructions change:

IX to IY and DD to FD

In the 8K monitor program the IY register pair always holds the value Hex. 4000 which corresponds to the base address of the system variables. The IX register pair is used in the 'display' routine as a store for a 'return address' and is therefore not available to the programmer unless programs contain their own 'display' routine.

An example from the 8K monitor program.

The individual system variables are often 'picked-up' using the instructions in this subgroup.

The following instruction line from part of the 'Clear screen routine' is used to 'pick-up' the system variable 'DF-SZ' which is the 'number of lines in lower part of screen' variable, 16418.

address	Hex. code	mnemonic	comment
0A1F	FD 46 22	LD B,(IY+22)	Load B with system variable 16418, DF-SZ

The BASIC programs that demonstrate these instructions are to be found on page 87.

Group 5. Instructions for loading locations in memory with data copies from registers, or with constants.

The instructions in this group perform operations that in general are the opposite of those described in group 4.

The instructions that form this group allow for the contents of the main registers to be copied to addressed locations in the memory, or for constants to be loaded into these locations.

The instructions can again be divided into three subgroups that each uses its own mode of addressing.

SUBGROUP A: INSTRUCTIONS USING 'ABSOLUTE ADDRESSING'

The following instructions form this subgroup:

mnemonic	**instruction** Hex.
LD (addr.),A	32 addr.
LD (addr.),HL	22 addr. (usual form)
	or
	ED 63 addr. (unusual form)
LD (addr.),BC	ED 43 addr.
LD (addr.),DE	ED 53 addr.
LD (addr.),IX	DD 22 addr.
LD (addr.),IY	FD 22 addr.
LD (addr.),SP	ED 73 addr.

Note that there is no instruction for directly storing a constant in an absolutely addressed memory location.

Once again the only register that can be copied, as a single register, is the A register. All of the other instructions involve the moving of two bytes of data. It is the contents of the LOW register of the register pair that is copied into the addressed location, and the contents of the HIGH register that goes into the following location.

The instructions that use 'absolute addressing' are very commonly used and the following example shows the whole of the 'CLEAR command routine'.

An example from the 8K monitor program

In the BASIC interpreter there are routines for each of the 31 BASIC commands. The CLEAR command has one of the simplest routines. When it is called, the routine loads a Hex.80 into the first location of the VARIABLE area. Then the address of the next location is stored as the E-LINE, STKBOT and STKEND system variables using instructions from this subgroup. The effect of these operations is to 'empty' the variable area and the workspace.

address	Hex. code	mnemonic	comment
1496	2A 10 40	LD HL,(4010)	Pick-up VARS
1499	36 80	LD (HL),+80	Enter Hex. 80
149B	23	INC HL	Move on one place.
149C	22 14 40	LD (4014),HL	E-LINE set.
149F	2A 14 40	LD HL,(4014)	Restore HL
14A2	22 1A 40	LD (401A),HL	STKBOT set.
14A5	22 1C 40	LD (401C),HL	STKEND set.
14A8	C9	RET	Return.

The 'LD (HL),+80' instruction is one of the instructions to be included in subgroup b. (see below). The 'INC HL' instruction is dealt with more fully on page 39.

Note that as the routine stands the instruction line containing the instruction 'LD HL,(4014)' is superfluous. It is present because the later instructions of the routine also form part of the RUN sequence, which does perform a CLEAR operation whenever it is called.

SUBGROUP B: INSTRUCTIONS USING 'INDIRECT ADDRESSING'

The instructions in this subgroup allow for a copy of the A register to be stored in a location addressed by the HL, BC or DE register pairs, and for the contents of the other main registers to be stored in a location addressed only by the HL register pair. The HL register pair can also point to a location that is to be loaded with a single byte constant.

It is also convenient to include in this subgroup three rather specialised instructions that allow the contents of HL, IX or IY register pairs to be exchanged with the last two bytes of data on the stack.

The actual instructions are:

mnemonic	**instruction** Hex.
LD (HL),A	77
LD (BC),A	02
LD (DE),A	12
LD (HL),H	74
LD (HL),L	75
LD (HL),B	70
LD (HL),C	71
LD (HL),D	72
LD (HL),E	73
LD (HL),+dd	36 dd
EX (SP),HL	E3
EX (SP),IX	DD E3
EX (SP),IY	FD E3

The instructions of this subgroup are often used to 'put-back' values following some sort of manipulation.

An example from the 8K monitor program

In the two earlier examples that use the 'Change all pointers' routine (pages 27 and 30) it was shown that the values of the pointers are 'picked-up' in turn, and then manipulated. The following instruction lines show how the results are 'put-back' using instructions from the present group under discussion.

address	Hex. code	mnemonic	comment
09C3	72	LD (HL),D	Put-back high byte.
09C4	2B	DEC HL	Decrease pointer
09C5	73	LD (HL),E	Put back low byte.

The instruction 'DEC HL' will be discussed in detail on page 42.

An example showing the use of the 'LD (HL),+dd' instruction is to be found on page 34 .

The three more specialised instructions that use 'indirect addressing' all involve the exchanging of the last two bytes on the stack with those bytes addressed by the HL, IX or IY register pair.

These instructions can be used in many instances, and the following example shows how the 'EX (SP),HL' instruction is used in the 8K monitor program.

An example from the 8K monitor program

The use of the stack by the BASIC interpreter is rather complicated as part of the stack area is used to hold the 'return line numbers' for the GOSUB-RETURN command and another part of the stack is used as the 'normal' working stack.

When a GOSUB command is executed the appropriate values are put into the GOSUB STACK, and when a RETURN command is executed, these values are collected. However the interpreter must check for the 'RETURN without GOSUB' error condition.

This is done using the 'EX (SP),HL' instruction. If there are no line numbers on the stack to be used as return line numbers then the high byte taken off the stack will be Hex. 3E.

Therefore the interpreter checks that the high byte on the stack is not Hex. 3E, if it is then error '7' is signalled.

The actual lines are:

address	Hex. code	mnemonic	comment
ØED9	E3	EX (SP),HL	Collect stack values.
EDA	7C	LD A,H	High byte to A register

. . . followed by test against Hex. 3E. Error 7 if equals Hex. 3E. Proceed to use line number, if not Hex. 3E.

SUBGROUP C: INSTRUCTIONS USING 'INDEXED ADDRESSING'.

The instructions in this subgroup allow the programmer to copy the contents of the main registers into an indexed addressed location, or to load a constant into the location.

The instructions are:

mnemonic	**instruction** Hex.
LD (IX+d), A	DD 77 d
LD (IX+d), H	DD 74 d
LD (IX+d),L	DD 75 d
LD (IX+d),B	DD 70 d
LD (IX+d),C	DD 71 d
LD (IX+d),D	DD 72 d
LD (IX+d),E	DD 73 d
LD (IX+d),+dd	DD 36 d dd

Note for the IY instruction change:

IX to IY and DD to FD

These instructions are used in the 'opposite' way to the instructions in group 4c.

The following example shows how a constant can be loaded into a location in a table as long as the 'base address' has been specified and the 'displacement' is known.

An example from the 8K monitor program

When the power is connected to the ZX-81, the computer enters 'slow' mode. The entry into this mode is not at random but arranged by the following line:

address	Hex. code	mnemonic
03FC	FD 36 3B 40	LD (403B),+40

This line from the 'initialisation' routine sets the flags of the system variable CD-FLAG, 16443, now with Bit 6 set (not bit 7) the computer will be in 'slow' mode until a change is made by the programmer.

The BASIC programs that demonstrate these instructions are to be found on page 89.

Group 6: The Addition Instructions.

The instructions in this group are the first instructions for handling arithmetic operations that are to be discussed. Later groups of instructions show how the Z80 can handle subtraction, comparison and logical operations.

The Addition instructions add in **absolute** binary arithmetic a specified number to the contents of a single register, a register pair or an indexed addressed location in memory.

The instructions in this group can be divided into three subgroups, with each subgroup having instructions with its own mnemonic type.

The subgroups are:

(a) The ADD instructions. These are straightforward addition instructions.

(b) The INC instructions. INCrementation just adds 1 to a specified number.

(c) The ADC instruction. These instructions are identical to the ADD instructions except that if the 'CARRY FLAG' is Set then the result of the addition is incremented as well.

Before the instructions in the group are discussed it is appropriate to describe the 'CARRY FLAG' in more detail.

The Carry Flag:

In reality the Carry flag is bit Ø of the F register. It is used in many different instances, but as an introduction to its function, the Carry flag can be considered to often act as an extra bit for the A register.

Normally the A register is described as having 8 bits, numbered Ø-7. Bit Ø is the right hand bit, the least significant bit, and bit 7 is the left hand bit, the most significant bit. However it is useful in many instances to consider the Carry flag as an 8th. bit to the register.

Note however that the Carry flag is just a bit in a different register and can be considered to be the 8th. bit of practically all the registers and not just the A register.

In respect of the Addition instructions the following statements can be made:

i. The ADD instructions ignore the state of the Carry flag before performing the addition, but Set (give value 1) the flag if the addition generates an extra binary digit, and Reset (give value Ø) if not.

i.e. Hex.

6Ø + 6C = CC Carry will be RESET
6Ø + AC = ØC Carry will be SET.

The addition of Hex. 6Ø + AC will give ØC when restricted to the 8 bits of a register, but the overflowing of the register leads to CARRY SET.

ii. The INC instructions **do not** take any notice of the Carry flag, and the result of an INC operation **does not** affect the Carry flag.

iii. The ADC instructions include the present state of the Carry flag in their additions, and leave the Carry flag Reset, or Set, depending on the final result of the ADC operation.

i.e. Hex.

Ø1 + 6Ø + 6C = CD Carry will be RESET
Ø1 + 6Ø = ØD Carry will be SET.
CARRY
FLAG
SET

With the Carry Flag Reset initially the results will be Hex.CC and ØC as before.

SUBGROUP A: THE ADD INSTRUCTIONS;

The instructions are:

mnemonic	**instruction** Hex	**mnemonic**	**instruction** Hex
ADD A,+dd	C6 dd	ADD HL,HL	29
ADD A,A	87	ADD HL,BC	09
ADD A,H	84	ADD HL,DE	19
ADD A,L	85	ADD HL,SP	39
ADD A,B	80	ADD IX,IX	DD 29
ADD A,C	81	ADD IX,BC	DD 09
ADD A,D	82	ADD IX,DE	DD 19
ADD A,E	83	ADD IX,SP	DD 39
ADD A,(HL)	86		
ADD A,(IX+d)	DD 86 d		

Note for the IY instructions change:

IX to IY and DD to FD

The ADD instructions are really very straightforward instructions, and are frequently used.

There are many instructions for forming the result of an addition in the A register, when dealing with single bytes, and the HL register pair when dealing with 2-byte numbers. However it is never possible to perform an addition and get the result directly in any of the other registers.

A good example of the use of an ADD instruction is given by the following example from the 'Keyboard decode routine'.

An example from the 8K monitor program.

There are 78 different keyboard codes generated by the 78 different keys on the keyboard of a ZX-81. In the 'keyboard decode routine' these codes are arranged so as to be in the range decimal 1-78, Hex. 01-4D. The appropriate characters for these codes are to be found in the 'character table' at Hex. 007E to 00CB inclusively.

For example the 'A' key will give a keyboard code of value '5', and the fifth character in the 'character table is decimal 38, Hex. 26, which is the ZX-81 code for the letter 'A'.

The 'character table' at Hex. 007E-00CB is for the 'L-mode'. The table at Hex. 00CC-00F2 is for 'F-mode' and the table at Hex. 00F3-0110 is for 'G-mode'

The 'keyword table' is at Hex.0111-01FB.

address	Hex. code	Mnemonic	comment
07D5	21 7D 00	LD HL,+007D	Set base address.
7D8	5F	LD E,A	Transfer to E.
7D9	19	ADD HL,DE	Form new address.

The routine operates with the keyboard code being originally in the A register. The base address is put into the HL register pair and the keyboard code is added to the base address. The D register holds zero throughout the routine. (see page 153 for the full 'Keyboard decode routine).

SUBGROUP B: THE INC INSTRUCTIONS

The instructions in this subgroup are used to add 1 to a specified 8 bit, or 16 bit number. The addition is in absolute binary arithmetic. The instructions neither take note of the Carry flag nor let the result of the addition affect the Carry flag.

The actual instructions are:

mnemonic	**instruction** Hex	**mnemonic**	**instruction** Hex
INC A	3C	INC HL	23
INC H	24	INC BC	03
INC L	2C	INC DE	13
INC B	04	INC SP	33
INC C	0C	INC IX	DD 23
INC D	14	INC IY	FD 23
INC E	1C		
INC (HL)	34		
INC (IX+d)	DD 34 d		
INC (IY+d)	FD 34 d		

Examples showing the use of INC instructions can be found on pages 31 and 34.

SUBGROUP C: THE ADC INSTRUCTIONS

The instructions in this subgroup enable the programmer to perform the addition between 8 bit, and 16 bit numbers, taking into account the present state of the Carry flag. The result of the addition is unchanged if the Carry flag is Reset, but incremented if the Carry flag is Set.

All the ADC instructions Set, or Reset, the Carry flag depending on the result of the final addition.

The actual instructions are:

mnemonic	**instruction** Hex	**mnemonic**	**instruction** Hex
ADC A,+dd	CE dd	ADC HL,HL	ED 6A
ADC A,A	8F	ADC HL,BC	ED 4A
ADC A,H	8C	ADC HL,DE	ED 5A
ADC A,L	8D	ADC HL,SP	ED 7A
ADC A,B	88		
ADC A,C	89		
ADC A,D	8A		
ADC A,E	8B		
ADC A,(HL)	8E		
ADC A,(IX+d)	DD 8E d		
ADC A,(IY+d)	FD 8E d		

The ADC instructions are not commonly used codes unless it is important to consider the 'carry' when using 'multiple precision arithmetic'. In such a case the 'carry' between bytes enables the programmer to hold numbers in more than two bytes.

The following example from the 8K monitor program however shows the Carry flag being used in the 'PRINT command routine'.

An example from the 8K monitor program

In the 'PRINT command routine' the characters for 'comma and semi-colon' have to be separated from the other characters. This is performed by arranging that: for the 'comma', the A register holds Hex. ØØ with Carry Reset and for the 'semi-colon' that the A register holds Hex.FF with Carry Set.

The routine then uses the ADC instruction. The effect of this being that the A register will hold Hex. ØØ for both the 'comma' and the 'semi-colon'.

address	Hex code	mnemonic	Comment
ØAD7	CE ØØ	ADC A,+ØØ	Just add the Carry.

The BASIC programs that demonstrate these instructions are to be found on page 92..

Group 7. The Subtraction Instructions.

The Subtraction instructions allow the programmer to subtract, using absolute binary arithmetic, a specified number from the contents of a single register, a register pair or an indexed addressed location in memory.

As with the Addition instructions, the Subtraction instructions can be divided into three subgroups, each with its own mnemonic.

There are:

(a) The SUB instructions that allow for single byte numbers to be subtracted from the contents of the A register.

(b) The DEC instructions that allow for the special case of subtracting 1 from a specified byte, or register pair.

(c) The SBC instructions that allow for the subtraction of the present value of the Carry flag as well as the normal subtraction.

In respect of the Subtraction Instructions the following statements can be made:

i. The SUB instructions are not affected by the present state of the Carry flag. However the Carry flag will be **Reset** if the result of the subtraction is **correct,** and **Set** if the result is **incorrect** because zero has been passed.

e.g.

in decimal 220 - 205	= 15 with CARRY RESET
in Hex. DC - CD	= OF
in decimal 208 - 208	= 0 with CARRY RESET
in Hex. DØ - DØ	= ØØ
in decimal 215 - 216	= 255 with CARRY SET
in Hex. D7 = D8	= FF

In effect 'Greater than' or 'Equals' gives RESET and 'Less than' gives SET.

ii. The DEC instructions do **not** take any notice of the Carry flag and the result of a DEC operation does not affect the Carry flag.

iii. The SBC instructions include the present state of the Carry flag into their subtractions. The result of the SBC operation also affects the Carry flag.

SUBGROUP A. THE SUB INSTRUCTIONS

The instructions in this subgroup are:

mnemonic	**instruction** Hex.
SUB +dd	D6 dd
SUB A	97
SUB H	94
SUB L	95
SUB B	9Ø
SUB C	91
SUB D	92
SUB E	93
SUB (HL)	96
SUB (IX+d)	DD 96 d
SUB (IY+d)	FD 96 d

The SUB instructions are the normal instructions for performing subtraction on single byte numbers, and are therefore commonly used instructions.

The following example shows the 'SUB +dd' instruction.

An example from the 8K monitor program.

In the previous example from the 'PRINT command routine' (page 40) it was said that the A register is made to hold: for the 'comma' the value Hex. ØØ with Carry Reset and for the 'semi-colon' the value Hex. FF with Carry Set.

These results are the direct result of the following line where the constant value Hex. 1A, decimal 26, is subtracted from the 'character codes' for 'comma' and 'semi-colon' held in the A register.

Address	Hex. code	mnemonic	comment
ØAD5	D6 1A	SUB +1A	Simply subtract '26'

For a 'comma'
In decimal 26 − 26 = Ø Carry Reset.
In Hex. 1A — 1A = ØØ
For a 'semi-colon'
In decimal 25 — 26 = 255 Carry Set
In Hex. 19 — 1A = FF

SUBGROUP B. THE DEC INSTRUCTIONS

The instructions in this subgroup are:

mnemonic	**instruction** Hex	**mnemonic**	**instruction** Hex
DEC A	3D	DEC HL	2B
DEC H	25	DEC BC	ØB
DEC L	2D	DEC DE	1B
DEC B	Ø5	DEC SP	3B
DEC C	ØD	DEC IX	DD 2B
DEC D	15	DEC IY	FD 2B
DEC E	1D		
DEC (HL)	35		
DEC (IX+d)	DD 35 d		
DEC (IX+d)	FD 35 d		

The DEC instructions are commonly used instructions when dealing with 'counters' and 'pointers'. The following example shows the 'DEC HL' instruction being used four times in quick succession in the 'Initialisation routine' where the HL register pair is being used as an address pointer.

An example from the 8K monitor program

Part of the 'Initialisation routine' involves loading the value Hex. 3E into the first location of the stack, (see also example on page 35) and then setting the Stack Pointer and the system variable ERR-SP, 16386, to their correct initial values.

Address	Hex. code	mnemonic	comment
Ø3E5	2A Ø4 4Ø	LD HL,(4ØØ4)	Fetch 'RAMTOP'
3E8	2B	DEC HL	Point to 1st loc.
3E9	36 3E	LD (HL),+3E	'return' marker
3EB	2B	DEC HL	Miss a location
3EC	F9	LD SP,HL	Store on SP
3ED	2B	DEC HL	Miss one
3EE	2B	DEC HL	Miss one
3EF	22 Ø2 4Ø	LD (4ØØ2),HL	Store as 'ERR-SP'

SUBGROUP C: THE SBC INSTRUCTIONS

The SBC instructions allow the programmer to perform subtraction taking into account the value of the Carry flag. There are SBC instructions for handling both single byte and 2-byte subtractions.

Note that the SBC instructions will not give the correct result of a subtraction operation unless the Carry flag is initially Reset. This can be a problem when dealing with 2-byte subtractions as there are no SUB instructions for handling 2-byte operations.

The usual way of 'clearing' (resetting) the Carry flag is to use the 'AND A' instruction (see page 45).

The instructions in this subgroup are:

mnemonic	**instruction** Hex	**mnemonic**	**instruction** Hex
SBC A,+dd	DE dd	SBC HL,HL	ED 62
SBC A,A	9F	SBC HL,BC	ED 42
SBC A,H	9C	SBC HL,DE	ED 52
SBC A,L	9D	SBC HL,SP	ED 72
SBC A,B	98		
SBC A,C	99		
SBC A,D	9A		
SBC A,E	9B		
SBC A,(HL)	9E		
SBC A,(IX+d)	DD 9E d		
SBC A,(IY+d)	FD 9E d		

Whereas the ADC instructions are uncommon the SBC instructions are widely used for many different reasons.

The following example shows the 'SBC HL,DE' instruction being used to compare two 16 bit numbers.

An example from the 8K monitor program

The 'print a whole BASIC line routine' is used to print an expanded BASIC line, replacing 'tokens' with appropriate 'words', i.e. character code 234 with the word REM. etc. It is also this routine that adds the 'inverse S' that shows a syntax error.

When the 'syntax checking routine' finds an error the appropriate address is stored as system variable 16408, X-PTR.

As the 'print a whole BASIC line routine' prints each character, a check is made to determine whether or not an 'inverse S' should be printed instead. The following lines show how the addresses are compared using a SBC instruction.

address	Hex. code	mnemonic	comment
Ø76D	ED 48 18 4Ø	LD BC,(4018)	Pick-up X-PTR
771	2A 16 4Ø	LD HL,(4016)	Pickup CH-ADD
774	A7	AND A	Clear Carry flag
775	ED 42	SBC HL,DE	Compare addresses.

. . . and print an 'inverse S' if the addresses match, this will be shown by the result being zero.

The BASIC programs that demonstrate these instructions are to be found on page 94.

Group 8: The Compare Instructions

The number of instructions in this group is very small, but the instructions are some of the most commonly used instructions in any machine code program.

The Compare instruction allows the programmer to compare the current contents of the A register with another byte of data. As usual this second byte of data can be a constant, the contents of a main register or the contents of an indexed addressed location in memory.

The Compare instructions perform a subtraction operation (without Carry) but the result is discarded after being used in the setting of the flags in the flag register.

The most important flag is the Carry flag and this is affected in the same manner as with the subtraction instructions. An operation in which the contents of the A register is 'Greater than' or 'Equal' to the second byte of data will RESET the Carry flag, whereas an operation with the contents of the A register being 'Less than' the second byte will SET the Carry flag.

The instructions are:

mnemonic	**instruction** Hex.
CP +dd	FE dd
CP A	BF
CP H	BC
CP L	BD
CP B	B8
CP C	B9
CP D	BA
CP E	BB
CP (HL)	BE
CP (IX+d)	DD BE d
CP (IY+d)	FD BE d

There are also block compare instructions but these are considered on page 72.

In the monitor program there are literally hundreds of comparisons and the following example illustrates the 'CP H' instruction, where the contents of the A register is compared to the existing contents of the H register.

An example from the 8K monitor program

As part of the initialisation of the ZX-81 there is a check made on the memory to determine just how much memory is available for use. This 'RAM check routine' tries to load the Hex. value Ø2 into every location between Hex. 7FFF and 4ØØØ. When a 16K RAM pack is fitted there will be a real location available for each attempt, but when less memory is fitted, the 'RAM check routine' will fail to load the actual location. Following the loading of the memory the routine continues by reading each location in turn, starting at Hex. 4ØØØ, until it finds a location that does not contain the value Hex. Ø2. The address of that location becomes RAMTOP.

A 'CP H' instruction is used in the following lines to test whether or not the contents of the H register has reached Hex. 3F as the routine goes down through the memory, entering the value Hex. Ø2.

address	Hex. code	mnemonic	comment
ØØØ2	01 FF 7F	LD BC,+7FFF	Set BC to top of possible RAM.
3CB	60	LD H,B	Transfer BC
3CC	69	LD L,C	to HL.
3CD	3F 3E	LD A,+3F	Load A.
3CF	36 Ø2	LD (HL),+Ø2	The Hex. Ø2.
3D1	2B	DEC HL	Next location.
3D2	BC	CP H	The comparison

. . . and back to Ø3CF if not 'equal'.

Note that the BC register is set to Hex. 7FFF in instruction line ØØØ2, as part of the power-on procedures.

The NEW command will use the RAM check routine to check up to RAMTOP by entering the routine at Ø3CB rather than ØØØØ as the value of RAMTOP may have been altered by the programmer in the meanwhile.

The BASIC program that demonstrates an instruction from this group is to be found on page 95.

Group 9: The Logical Instructions

There are instructions for performing the logical operations of AND, OR and XOR. These operations are performed between the contents of the A register and a specified byte.

Each subgroup will now be discussed in turn:

SUBGROUP A: THE AND INSTRUCTIONS

The logical operation AND allows the programmer to 'AND' the individual bits of the A register with the corresponding bits of a specified byte. The result of the 8 operations is returned to the programmer in the A register.

The AND operation specifies that the resultant bit of the test between two bits will be Set only if both of the bits under test are themselves Set. Otherwise the resultant bit will be Reset.

e.g.	1 0 1 0 1 0 1 0		AA
in binary	AND		AND
	1 1 0 0 0 0 0 0	in Hex.	CO
	results in		results in
	1 0 0 0 0 0 0 0		80

The effect of the AND operation can be summarised as being an operation that KEEPS certain bits in the A register, or MASKS OFF the other bits, depending on the point of view that is taken.

In the example above the operation of 'AND + CO' will result in the KEEPing only of the Bit 6 & the Bit 7 in the A register; or the MASKing OFF of Bits 0-5 inclusive.

The AND instructions also clear the Carry flag, and hence 'AND A' is often used as the 'clear carry flag' instruction.

The instructions in this subgroup are:

mnemonic	**instruction** Hex.
AND +dd	E6 dd
AND A	A7
AND H	A4
AND L	A5
AND B	A0
AND C	A1
AND D	A2
AND E	A3
AND (HL)	A6
AND (IX+d)	DD A6 d
AND (IY+d)	FD A6 d

The following example shows the 'AND +dd' instruction being used to MASK OFF two bits.

An example from the 8K monitor program.

The 'Keyword table' at 0111 - 01FB holds the expanded forms of all keywords.

e.g. the keyword LIST is held as:

Hex. 31	in location	01AF
Hex. 2E	in location	01B0
Hex. 38	in location	01B1
Hex. B9	in location	01B2

where the character code for 'L' is Hex. 31, for 'I' is Hex. 2E, for 'S' is Hex. 38, and the character code Hex. B9 is an 'inverted T' that marks the end of the word.

In the 'print keyword routine' are the instructions that find the appropriate word for a particular keyword code. The letters of the word are then printed on the T.V. display in the following manner:

Each letter is 'ANDed' with Hex. 3F. This removes nothing from the normal letters but removes the 'inverting' from the last letter of the word.

The letter is then printed.

The letter is fetched again and tested at this point to find out whether or not it is a 'last letter'.

The test is performed by 'doubling'. The double of an ordinary number will give Carry Reset, whereas the double of an inverse character will give Carry Set.

The instruction lines are:

address	Hex. code	mnemonic	comment
0959	0A	LD A,(BC)	Pick-up letter.
95A	E6 3F	AND +3F	Mask-off bits 6 & 7.
. . . print the character.			
95D	0A	LD A,(BC)	Fetch letter again.
95E	03	INC BC	Move pointer forward.
95F	87	ADD A,A	Test by doubling
. . . return to 0959 if Carry Reset.			

SUBGROUP B: THE OR INSTRUCTIONS

The logical operation OR allows the programmer to 'OR' the individual bits of the A register with the corresponding bits of a specified byte.

The OR operation specifies that the resultant bit of the test between two bits will be Set only if either of the bits under test are themselves Set. Otherwise the bit will be Reset.

e.g.	1 0 1 0 1 0 1 0		AA
in binary	OR	in Hex.	OR
	1 1 0 0 0 0 0 0		CO
	results in		results in
	1 1 1 0 1 0 1 0		EA

The OR instructions are not very commonly used instructions and indeed they can usually be avoided completely if it is wished. However there are certain times when the use of an OR operation is the easiest way of performing the operation.

Whereas the effect of the AND operation was to MASK OFF certain bits, the OR operation SETS certain bits.

The OR instructions are:

mnemonic	**instruction** Hex.
OR +dd	F6 dd
OR A	B7
OR H	B4
OR L	B5
OR B	B0
OR C	B1
OR D	B2
OR E	B3
OR (HL)	B6
OR (IX+d)	DD B6 d
OR (IY+d)	FD B6 d

The following example shows the 'OR +dd' instruction being used to SET bits 5, 6 and 7 of the A register.

An example from the 8K monitor program.

The 'keyboard decode routine' of the 8K monitor program allows for the keyboard to be scanned eight times. On each scan the address bus holds a different address so in effect the keyboard is scanned from 'eight different directions', one after another.

There are only five data lines coming from the keyboard and it is necessary to Set the three unused lines so that their contents are predictable.

The following line performs this operation:

address	Hex. code	mnemonic	comment
Ø2C5	F6 EØ	OR +EØ	Sets bits 5, 6 & 7 so as to make them predictable

(see page 153 for full 'Keyboard decode routine')

SUBGROUP C: THE XOR INSTRUCTIONS

The logical operation XOR allows the programmer to 'XOR' the individual bits of the A register with the corresponding bits of a specified byte.

The XOR operation specifies that the resultant bit of the test between bits will be Set only if either, but not both, of the bits under test are themselves Set. Otherwise the bit will be Reset.

e.g.	1 Ø 1 Ø 1 Ø 1 Ø		AA
in binary	XOR	in Hex.	XOR
	1 1 Ø Ø Ø Ø Ø Ø		CØ
	results in		results in
	Ø 1 1 Ø 1 Ø 1 Ø		6A

Once again the instructions in this group are uncommon instructions, and their use can usually be completely avoided. However the 'XOR A' instruction can be useful as it has the effect of clearing the A register in just a single byte of machine code. It does however also clear the Carry flag which may not be always helpful.

The actual instructions are:

mnemonic	**instruction** Hex.
XOR +dd	EE dd
XOR A	AF
XOR H	AC
XOR L	AD
XOR B	A8
XOR C	A9
XOR D	AA
XOR E	AB
XOR (HL)	AE
XOR (IX+d)	DD AE d
XOR (IY+d)	FD AE d

Apart from the use of the 'XOR A' instruction to clear the A register, the use of the XOR instructions is always complicated programming. The following example from the monitor program shows how string variables are marked as string variables.

An example from the 8K monitor program.

A string variable is marked as being a string variable by having bit 7 RESET, bit 6 SET and bit 5 RESET. This is achieved in the 'Assignment of a string variable routine' as follows:

The address of the single letter of the name is loaded into the HL register pair.

Then by using a 'XOR HL' instruction the character code for the letter is XORed against Hex. 60. This has the effect of Resetting bit 7, Setting bit 6 and inverting bit 5 which will always give a Reset value.

The lines are:

address	Hex. code	mnemonic	comment
13C4	3E 60	LD A,+60	Load A with Hex. 60
13C6	2A 12 40	LD HL,(4012)	Pick-up address
13C9	AE	XOR (HL)	Change bits.

The BASIC programs that demonstrate these instructions are to be found on page 96.

Group 10. The Jump Instructions.

In the Z80 instruction set there are 17 instructions that allow the programmer to make jumps from one part of his program to another.

The instructions can be split into seven subgroups depending on the type of jump involved.

Each subgroup will now be discussed in turn:

SUBGROUP A: THE ABSOLUTE JUMP INSTRUCTION

The single instruction in this subgroup is the simplest of all the Jump Instructions. The instruction uses absolute addressing to specify the address to which the jump is to be made.

The instruction is:

mnemonic	**instruction** Hex.
JP addr.	C3 addr.

This instruction is directly equivalent to the BASIC 'GOTO' command.

The execution of this command does not take any notice of the present state of any of the flags and does not affect any of the flags. The result of the instruction is simply to load the Program Counter register pair with the specified address.

The instruction is used simply to move from one block of code to another, past subroutines and tables in particular.

An example from the 8K monitor program.

The first occurrence of an absolute jump instruction comes in the 'start routine' at 0005 when a jump is made past various subroutines and character tables.

address	Hex. code	mnemonic	comment
0005	C3 CB 03	JP 03CB	Make a jump.
03CB	60	LDH,B	Jumps to here.

SUBGROUP B: JUMP INSTRUCTIONS THAT USE INDIRECT ADDRESSING

There are three instructions that enable jumps to be made to indirect addressed locations. Surprisingly the register pairs are the HL, IX and IY register pairs rather than the usual set of HL, BC and DE register pairs.

The instructions are:

mnemonic	**instruction** Hex.
JP (HL)	E9
JP (IX)	DD E9
JP (IY)	FD E9

The effect of the execution of these instructions is to load the Program Counter with the specified address. The flags are not affected.

An example from the 8K monitor program.

The 8K monitor program uses the 'JP (HL)' instruction in its 'display routine' in a very interesting way. In this routine bit 15 of the HL register pair is Set, (by using a SET 7,H instruction) and then a jump is made to this location. In reality such a jump will take the Z80 out of the monitor program but because of a special 'switch' in the circuitry of the ZX-81 the Z80 will start executing NOP instructions until an 'interrupt' occurs. This 'interrupt' will always occur after 32 NOP instructions as this is the length of a display line.

address	Hex. code	mnemonic	comment
ØØ44	E9	JP (HL)	Start executing NOP's.

SUBGROUP C: THE RELATIVE JUMP INSTRUCTION.

The single instruction in this subgroup, the 'JR e' instruction allows the programmer to make jumps to locations that are within 127 locations forward and 128 locations back from the current position.

The 'JR e' instruction is a very useful instruction as it allows the programmer to make short jumps using just two bytes of code, whereas a 'JP addr.' instruction uses three bytes of code.

The following diagram shows how the different values of 'e' are used to specify jumps of varying length. The value of 'e' is always considered to be in 2's complement arithmetic.

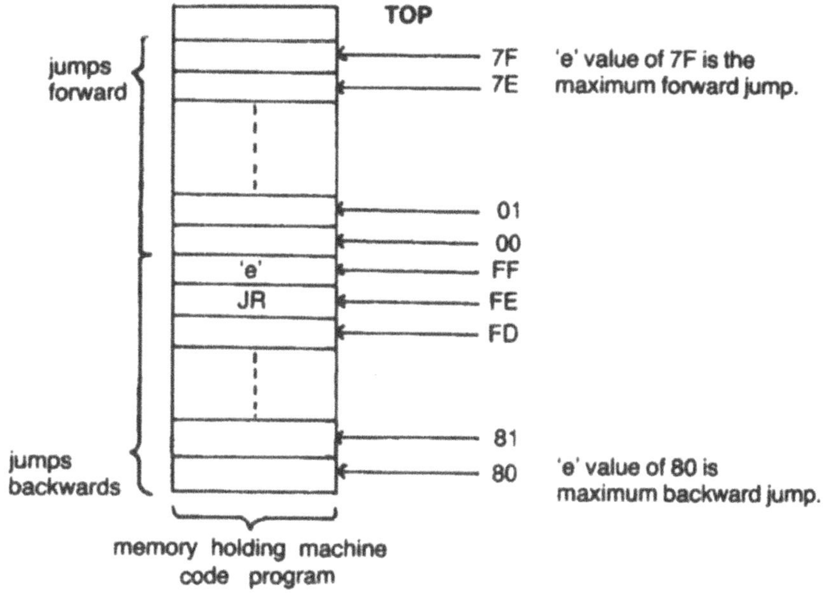

Diagram 8: To show how different values of 'e' cause jumps to different locations.

All programmers use their own 'rules of thumb' for calculating the required values of 'e' when they are writing machine code. The author considers that for forward jumps the value of 'e' represents the number of bytes that are 'jumped over'. Whereas the more difficult backwards jumps are obtained by counting backwards, in Hex., the 'e' location being called 'FF', until the required location has been reached. The Hex. value assigned to this location is then the correct value for 'e'.

The instruction is:

mnemonic	**instruction** Hex.
JR e	18 e

In the monitor program the 'JR e' instruction is used many times. The following example comes from the 'error handling routine':

An example from the 8K monitor program.

The 'error handling routine' begins at 0008 but it then has to jump past some other subroutines. As the required jump is only over Hex.46 bytes of code a 'JR e' instruction can be used.

address	Hex.code	mnemonic	comment
000E	18 46	JR 0056	Jump past Hex.46 bytes
0056			and continue from location 0056.

SUBGROUP D: JUMP INSTRUCTIONS CONDITIONAL ON THE CARRY FLAG.

There are four instructions that allow the programmer to make jumps that are conditional on the state of the Carry flag. These instructions are equivalent to the BASIC lines;

IF C THEN GOTO. . . .

and

IF NOT C THEN GOTO. . . . .

Two of the instructions allow for absolute addressed jumps, whilst the other two instructions allow for relative jumps.

It is sensible at this point to describe the Carry flag in detail before proceeding further.

The Carry Flag:

This flag is bit 0 of the flag register and has already been mentioned because of its use in arithmetic operations.

The following statements can be made concerning the Carry flag:

i. ALL ADD and ADC instructions will Set or Reset the Carry flag, depending on whether or not the result overflows the space allowed for it.

ii. All SUB, SBC and CP instructions will Reset the Carry flag if the result is 'correct', and Set the flag if zero has been passed.

iii. All AND, OR and XOR operations Reset the Carry flag.

iv. Rotation instructions (see page 66) affect the Carry flag.

v. LD instructions **do not** affect the Carry flag.

There are two instructions for handling the Carry flag and they are:

mnemonic	**instruction** Hex.
SCF	37
CCF	3F

The 'SET CARRY FLAG' instruction (SCF) simply Sets the Carry flag.

The 'COMPLEMENT CARRY FLAG' instruction (CCF) changes the value of the Carry flag from 0 to 1, or vice versa. Note that 'AND A' is the usual instruction for clearing the Carry flag.

The actual instructions in this subgroup are:

mnemonic	**instruction** Hex.	**Comment**
JP NC,addr.	D2 addr.	Jump on Carry flag Reset.
JP C,addr.	DA addr.	Jump on Carry flag Set.
JR NC,e	30 e	Jump on Carry flag Reset.
JR C,e	38 e	Jump on Carry flag Set.

In any machine code program the instructions that allow for jumps to be made conditional on the state of the Carry flag are very common. The following example from the monitor program shows this type of instruction being used three times in a short machine code routine.

An example from the 8K monitor program.

In the 'test PRINT AT parameters routine' the parameters are tested to see that they are in the correct range. The line parameter must be less than 22, and the column parameter must be less than 32.

The testing of the parameters is done by the following lines:

address	**Hex. code**	**mnemonic**	**comment**
08F5	3E 17	LD A,+17	Test line
8F7	90	SUB B	parameter against hex. 17.
8F8	38 0B	JR C,0905	Error B if fails.
8FA	FD BE 22	CP (4022)	Test v. lower screen.
8FD	DA 35 08	JP C,0835	Error 5 if fails.
900	3C	INC A	Store back
901	47	LD B,A	in B.

902	3E 1F	LD A,+1F	Test column
904	91	SUB C	parameter
			against Hex. 1F.
905	DA AD 0E	JP C,0EAD	Error B if fails.
908	C6 02	ADD A,+02	Store back in
90A	4F	LD C,A	C.

The conditional jump instructions are used to make jumps to the routines that handle the different errors.

SUBGROUP E: JUMP INSTRUCTIONS CONDITIONAL ON THE ZERO FLAG.

There are again four instructions that allow the programmer to make jumps that are conditional on the state of the Zero flag. These instructions are equivalent to the BASIC lines:

IF Z THEN GOTO ......

and

IF NOT Z THEN GOTO ......

Two of the instructions allow for absolute addressed jumps, whilst the other two instructions allow for relative jumps.

It is sensible at this point to describe the Zero flag in detail before proceeding further.

The Zero Flag:

This flag is bit 6 of the flag register and it is Set if the result of an operation is zero, and Reset if otherwise.

i.e. For operations in which the result is put into the A register, the Zero flag will be Set if the A register holds zero, otherwise the Zero flag will be Reset.

The following statements can be made concerning the **Zero flag:**

i. ALL ADD, INC, ADC, SUB, DEC, SBC, CP, AND, OR and XOR instructions using single registers, and ADC and SBC instructions using register pairs, affect the Zero flag.

ii. Rotation instructions (see page 66) affect the Zero flag.

iii. Bit testing instructions (see page 69) affect the Zero flag.

iv. LD instructions (except 'LD A,I' and 'LD A,R') **do not** affect the Zero flag.

v. Block searching instructions (see page 72) use the Zero flag.

vi. There are no instructions for explicitly handling the Zero flag.

The actual instructions in this subgroup are:

mnemonic	**instruction** hex.	**comment**
JP NZ,addr.	C2 addr.	Jump on Zero flag Reset.
JP Z,addr	CA addr.	Jump on Zero flag Set.
JR NZ,e	2Ø e	Jump on Zero flag Reset.
JR Z,e	28 e	Jump on Zero flag Set.

Once again the instructions in this subgroup are very common instructions. The following example shows the instructions being used in the 'PRINT command routine'.

An example from the 8K monitor program.

In the 'PRINT command routine' the different codes of a BASIC line are tested to see whether they are NEWLINE characters, commas or semi-colons, keywords or the word 'AT'. These tests are performed by using conditional jump instructions.

The lines involved are:

address	**Hex.code**	**mnemonic**	**comment**
ØACF	7E	LD A,(HL)	Pick-up code.
ADØ	FE 76	CP +76	Is it NEWLINE?
AD2	CA 84 OB	JP Z,ØB84	Jump if it is.
AD5	D6 1A	SUB +1A	Make 'comma
AD7	CEØØ	ADC A,+ØØ	and semi-colon' the same.
AD9	28 69	JR Z,ØB44	Jump if ',' or ';'.
ADB	FE A7	CP +A7	Test if 'AT'.
ADD	2Ø 1B	JR NZ,ØAFA	Jump if not 'AT'.
ADF	. . . . . .		Proceed with 'AT'.

SUBGROUP F: JUMP INSTRUCTIONS CONDITIONAL ON THE SIGN FLAG.

There are two instructions that allow the programmer to make jumps that are conditional on the state of the Sign flag. Both instructions use absolute addressing.

These instructions are equivalent to the BASIC lines:

IF S>=0 AND S<=127 THEN GOTO . . . . .

and

IF S>=128 AND S<=255 THEN GOTO . . . . .

It is sensible at this point to describe the Sign flag in detail before proceeding further.

The Sign Flag:

The Sign flag is bit 7 of the flag register and it shows whether a result is negative or positive with respect to 2's complement arithmetic. As bit 7 of a single register and bit 15 of a register pair are sign bits, it follows that the Sign flag is just a copy of the appropriate sign bit of a register or register pair. The following statements can be made concerning the Sign flag.

i. All ADD, INC, ADC, SUB, DEC, SBC, CP, AND, OR and XOR instructions using single registers, and ADC and SBC instructions using register pairs, affect the Sign flag.

ii. Rotation instructions (see page 66) affect the Sign flag.

iii. LD instructions (except 'LD A,I' and 'LD A,R') **do not** affect the Sign flag.

iv. Block searching instructions (see page 72) use the Sign flag.

v. There are no instructions for explicitly handling the Sign flag.

The actual instructions in this subgroup are:

mnemonic	**instruction** Hex	**Comment**
JP P,addr.	F2 addr.	Jump if result positive.
JP M,addr.	FA addr.	Jump if result negative (minus).

Both these instructions are rather uncommon and their use can usually be avoided. However the 'JP M,addr.' instruction can be used to test bit 6 as in the following example.

An example from the 8K monitor program.

In the 'next variable or BASIC line routine 'the different classes of variables are distinguished from each other by the different configurations of bits 5, 6 and 7.

The bit 5 is tested using a 'BIT 5' instruction (see page 69).

The bit 7 is tested by 'doubling' and looking at the resultant state of the Carry flag.

The bit 6 is tested by 'doubling' and then using the 'JP M,addr.' instruction to test the state of the new bit 7.

The code for the initial letter of the variable is held in the A register, and the following lines perform the test.

address	Hex.code	mnemonic	comment
09FC	87	ADD A,A	'Double'
9FD	FA 01 0A	JP M,0A01	Jump if bit 7 Set.

SUBGROUP G: JUMP INSTRUCTIONS CONDITIONAL ON THE OVERFLOW/PARITY FLAG.

There are two instructions that allow the programmer to make jumps that are conditional on the state of the Overflow/Parity flag. Both instructions use absolute addressing.

It is sensible at this point to describe the Overflow/Parity flag before proceeding further.

The Overflow/Parity Flag:

This flag is bit 2 of the flag register and as the name implies it is a dual purpose flag. Certain groups of instructions use the flag to denote 'overflow' whilst other groups of instructions use the flag to store the result of a 'parity test'.

The concept of 'overflow' does not relate to there being insufficient room, as is tested by the Carry flag, but rather to the test as to whether the result of an operation in **2's complement arithmetic** is giving a correct or incorrect answer.

e.g. Consider: Hex. 14 ADD Hex. 6F where the result will be Hex. 83. This result is correct when dealing with absolute binary arithmetic, but it is **incorrect** in 2's complement arithmetic.

The decimal interpretation of this operation shows the error condition clearly.

Hex. 14 + 6F = 83 INCORRECT
Decimal 20 + 111 = -125 INCORRECT

This operation will therefore SET the Overflow/Parity flag.

Overflow can also occur in subtraction operations. Consider:

Hex. 83 – 6F = 14 INCORRECT
Decimal –125 – 111 = 20 INCORRECT

when the Overflow/Parity flag will again be Set. The concept of 'parity' refers to the testing of the bits in a byte to determine whether there is an even, or odd, number of bits Set.

e.g. Consider the byte: 01010101 in which there is an even number of bits which are Set. (4). Therefore the Overflow/Parity flag will be SET.

Consider next the byte: 10101110 in which there is an odd number of bits which are Set. (5). Therefore the Overflow/Parity flag will be RESET.

The following statements can be made concerning the Overflow/Parity flag.

i. All ADD, ADC, SUB, SBC and CP operations are tested for 'overflow'.

ii. All AND, OR and XOR operations are tested for 'parity'.

iii. The results of Rotation operations are tested for 'parity' (see page 66).

iv. The INC instruction will Set the flag if the result is Hex.80.

v. The DEC instruction will SET the flag if the result is Hex.7F.

vi. The block instructions (see page 72) use the Overflow/Parity flag.

vii. There are no instructions for explicitly handling the Overflow/Parity flag.

The actual instructions in this subgroup are:

mnemonic	instruction Hex.	**Comment**
JP PØ,addr.	E2 addr.	Jump if parity odd, or no overflow. (flag Reset)
JP PE,addr.	EA addr.	Jump if parity even, or overflow. (flag Set)

The two instructions are both rarely used instructions except when 'parity' is being tested. But in the ZX-81 system there is no parity check on incoming or outgoing data.

The following example shows the flag being used to test for 'overflow'.

An example from the 8K monitor program.

In the 'character code sorting routine' the different types of character codes have to be separated. The codes for the simple characters are in the range Hex.ØØ – 3F, whereas the codes for the editing commands are in the range Hex.7Ø – 79.

The different types are separated using a 'JP PE,addr.' instruction.

The lines are:

address	Hex.code	mnemonic	comment
Ø515	7E	LD A,(HL)	Pick-up code.
516	FE FØ	CP +FØ	Compare it to +FØ
518	EA 2D Ø5	JP PE,Ø52D	Jump on overflow.

51B . . . . . proceed with simple characters.

The routine works by comparing the character code, that it has collected from the main character table, to the constant Hex. FO. All the codes for simple characters will lead to there being '**no** overflow'. e.g. The letter 'Z' with code. Hex. 3F will give:
3F – FO = 4F

or in decimal, 63 –(–16) = 79
which is CORRECT and Resets the flag.

The codes for the editing commands will however give 'overflow'.

e.g. The code for RUBOUT is Hex.77, and
77 – FØ = 87
or in decimal 119 –(–16) = –121

which is INCORRECT and Sets the flag, and thereby leads to the jump occurring.

The BASIC programs that demonstrate these instructions are to be found on page 97.

Group 11: The 'DJNZ e' instruction.

This single instruction is one of the most useful instructions in the Z80 instruction set.

The 'DJNZ e' instruction is an instruction that Decrements the B register and does a relative Jump if the resultant value in the B register is Not Zero. The instruction has many similarities to the BASIC command NEXT.

In a BASIC program the FOR command is used to set up a loop variable and the NEXT command delimits the working part of the loop.

When a 'DJNZ e' instruction is used, the B register becomes the loop variable and the 'DJNZ e' instruction is itself used as the equivalent of the NEXT command.

The following example shows these similarities:

BASIC program	**Machine Code language program**	
FOR I=1Ø TO 1 STEP -1	loop variable initialised.	LD B,+1Ø
LET . . .	the working	LD . . .
LET . . .	part.	LD . . .
NEXT I	the delimiter.	DJNZ e

Note how the loop in the example uses 'STEP -1', this is because the 'DJNZ e' instruction is always a Decrementing instruction.

It is also important to note that the value of 'e' is the displacement value that will take the 'loop' back to the start of the 'working part' of the program. Interestingly it is a common mistake to 'jump back' to the loop variable and hence create an endless machine code loop.

The above example shows the normal use of the 'DJNZ e' instruction, but it is also possible to use it with 'forward jumps' but that is indeed 'complicated programming' and perhaps is best avoided.

The actual instruction is:

mnemonic	**instruction** Hex.
DJNZ e	10 e

This instruction is very commonly used and the following example from the 'initialisation routine' is just one of many possible examples.

An example from the 8K monitor program.

When the ZX-81 is switched on, the 'initialisation routine' has to construct a display file that consists of 25 NEWLINE characters. (Hex.76, decimal 118). This operation is performed using a 'DJNZ e' instruction.

Initially the HL register pair holds the current D-FILE and then the B register is loaded with the loop variable Hex.19, decimal 25.

The working part of the loop consists of a 'LD (HL),+76' instruction, that enters the NEWLINE character and a 'INC HL' that moves the pointer forward one location.

The 'DJNZ e' instruction is then used to decrement the B register and jump back to repeat the 'working part' if the loop variable is not zero.

The lines involved are as follows:

address	**Hex. code**	**Label**	**mnemonic**	**comment**
Ø4Ø6	Ø6 19		LD B,-19	Loop variable.
4Ø8	36 76	W-Part	LD (HL),+76	NEWLINE
4ØA	23		INC HL	Forward one.
4ØB	1Ø FB		DJNZ W-Part	Back to 'W-Part'

Note in the above lines how a 'Label' field has been used to mark the start of the working part with the label 'W-Part', and how the mnemonic for the 'DJNZ e' instruction has been written as 'DJNZ W-Part'. The use of labels will be found frequently from now on.

Note also how the value of 'e' is given as Hex.FB in location 040C. The author finds that the best way of checking that this value of 'e' is correct, is to count back from 040C. The location 040B being equivalent to 'FC', the location 040A being equivalent to 'FD', and so on until the start of the correct loop will have the value 'FF'. In the example the value of 'e' being 'FB' is correct as it loops back to the start of the instruction line labelled W-Part.

A BASIC program that demonstrates the 'DJNZ e' instruction is to be found on page 99.

Group 12: The 'Stack' instructions.

The stack is used extensively in most machine code programs and therefore there are many instructions for handling the data on the stack.

The instructions can be divided into two subgroups. The first subgroup contains the instructions that can be used by the programmer to handle data that is stored temporarily on the stack. The second group of instructions contains those instructions that use the stack themselves.

SUBGROUP A: THE PUSH AND POP INSTRUCTIONS.

These instructions allow the programmer to put, 'PUSH', two bytes of data on to the stack, and to remove, 'POP', two bytes of data off the stack.

The two bytes of data in a PUSH operation must be copied from a specified register pair, and in a POP operation must be copied into a specified register pair.

When a PUSH operation is performed the Stack Pointer is first decremented, a copy of the high register byte is made and stored in the location addressed by the Stack Pointer. Then the Stack Pointer is decremented a second time and a copy of the low register byte is made and this byte is likewise stored in the location addressed by the Stack Pointer. The opposite actions are taken during a POP operation.

The actual instructions in this subgroup are:

mnemonic instruction	Hex.	**mnemonic instruction**	Hex.
PUSH AF	F5	POP AF	F1
PUSH HL	E5	POP HL	E1
PUSH BC	C5	POP BC	C1
PUSH DE	D5	POP DE	D1
PUSH IX	DD E5	POP IX	DD E1
PUSH IY	FD E5	POP IY	FD E1

It is important to realise that when two bytes are PUSHed on to the Stack, that no record is kept in any way that shows where the data came from. Two bytes coming from the HL register pair may therefore be POPed into the BC register pair or indeed any register pair.

The storing of data on the stack as a temporary measure is illustrated in the following example.

An example from the 8K monitor program.

In the 'display routine' the current values held in the four main register pairs are saved on the stack, when 'slow mode' interrupts the normal execution of a program. Later the contents of the registers are restored.

Note how the 'saving' is done in a certain order, and how the 'restoring' is done in the reverse order so as to get the correct contents restored.

address	Hex.code	Label	mnemonic	comment
0220	F5	Saving	PUSH AF	Save copies of
221	C5		PUSH BC	the main
222	D5		PUSH DE	register pairs
223	E5		PUSH HL	on the stack.

02A4	E1	Restoring	POP HL	Restore the
2A5	D1		POP DE	contents of the
2A6	C1		POP BC	main register
2A7	F1		POP AF	pairs.

SUBGROUP B: THE CALL, RET AND RST INSTRUCTIONS.

The execution of all these instructions results in either addresses being put on the stack, or return addresses being removed from the stack.

The CALL and the RST instructions are directly equivalent to the BASIC 'GOSUB' command, and the RET instructions are equivalent to the 'RETURN' command. There are however several conditional CALL and RET instructions that have no direct equivalent in BASIC.

Each of these three subgroups will now be discussed in turn.

SUBGROUP A: THE CALL INSTRUCTIONS.

The CALL instructions are used to enter subroutines, therefore the return address has to be saved. This is done by the high byte of the Program Counter being copied and saved on the stack and then the low byte being copied and saved. The Stack Pointer is decremented before each byte is stored. The Program Counter is then loaded with copies of the two bytes of data that follow the CALL instruction proper. As usual the low byte is the first of these bytes of data and the high byte the second byte of data.

There are instructions that allow for the execution of subroutines to be made conditional on both states of the four major flags.

The actual instructions in this subgroup are:

mnemonic	**instruction** Hex.	**Comment**
CALL addr.	CD addr.	Unconditional GOSUB.
CALL C,addr.	DC addr.	GOSUB, Carry flag Set.
CALL NC,addr.	D4 addr.	GOSUB, Carry flag Reset.
CALL Z,addr	CC addr.	GOSUB, Zero flag Set.
CALL NZ,addr.	C4 addr.	GOSUB Zero flag Reset.
CALL M,addr.	FC addr.	GOSUB, Sign flag Set.
CALL P,addr.	F4 addr.	GOSUB, Sign flag Reset.
CALL PE,addr.	EC addr.	GOSUB, O/P flag Set.
CALL PO,addr.	E4 addr.	GOSUB, O/P flag Reset.

In any large machine code program there will be a large number of subroutines and hence CALL instructions are fairly common instructions. However conditional CALL instructions are always fairly uncommon.

The following example shows an interesting feature of the ZX-81 system.

An example from the 8K monitor program.

In a ZX-81 system that has less than 3¼K of available RAM the display file is 'collapsed'. That is to say that an empty display line will consist of only a NEWLINE character, and partially completed lines will have a NEWLINE character positioned after the last defined character of the line.

In a ZX-81 system with more than 3¼K the display file is usually fully defined, with 24 lines of 32 characters.

In the 'build up an edit line routine' it is necessary to test for the size of available RAM so that the display file can be expanded to hold the edit line, if necessary.

The actual lines involved are:

address	**Hex.code**	**mnemonic**	**comment**
Ø4B7	3A Ø5 4Ø	LD A,(4ØØ5)	High byte of RAMTOP
4BA	FE 4D	CP +4D	Test for 3¼K of RAM
4BC	DC 5D ØA	CALL C,ØA5D	GOSUB if less RAM.

SUBGROUP B: THE RST INSTRUCTIONS.

In the Z80 instruction set there are eight RST instructions. These instructions are in effect CALL instructions but the address of the subroutine does not have to be specified as it is predetermined.

e.g. The instruction C7 which has the mnemonic 'RST 0000' is a single byte instruction that has the predetermined address of 0000. The instruction performs in just one byte the operation that would take three bytes using a 'CALL addr.' instruction.

The eight RST instructions all have predetermined addresses that are in the range 0000–0038.

The actual instructions are:

mnemonic	**instruction** Hex.	**comment**
RST 0000	C7	'GOSUB 0000'
RST 0008	CF	'GOSUB 0008'
RST 0010	D7	'GOSUB 0010'
RST 0018	DF	'GOSUB 0018'
RST 0020	E7	'GOSUB 0020'
RST 0028	EF	'GOSUB 0028'
RST 0030	F7	'GOSUB 0030'
RST 0038	FF	'GOSUB 0038'

In a Z80 machine code language program that is started at location 0000 it is therefore a normal feature to find that there are several commonly used subroutines starting at these predetermined addresses. The 8K monitor program is typical in this matter and the following list shows the routines called by the different RST instructions.

RST	**Routine**
0000	Start routine.
0008	Error handling routine.
0010	Print character routine.
0018	Collect present character routine.
0020	Collect next character routine.
0028	Calculator routine.
0030	Make room in memory routine.
0038	Display interrupt routine.

The following example shows how the 'print character routine' is called by using 'RST 0010'.

An example from the 8K monitor program.

The following extract from the 'print keyword routine' shows that the code of the character to be printed has to be present in the A register before the 'RST 0010' instruction can be used.

address	Hex. code	mnemonic	comment
0957	AF	XOR A	Zero A register.
958	D7	RST 0010	Print a zero – a space.
959	0A	LD A,(BC)	Pick-up character.
95A	E6 3F	AND +3F	mask character.
95C	D7	RST 0010	Print the character
95D................			

SUBGROUP C: THE RET INSTRUCTIONS

This subgroup consists of a simple RET instruction and eight conditional instructions. The main use of all these instructions is to act as an end marker of a subroutine and cause a return to the main program.

It is important to realise that a RET instruction does no more than take a return address from the stack and put it into the Program Counter. The first byte copied from the stack forms the low byte of the Program Counter and the second byte forms the high byte of the Program Counter. The Stack Pointer is incremented twice during any RET instruction.

The actual instructions are:

mnemonic	instruction Hex.	comment
RET	C9	The simple RETURN.
RET C	D8	RETURN if Carry flag Set.
RET NC	D0	RETURN if Carry flag Reset.
RET Z	C8	RETURN if Zero flag Set.
RET NZ	C0	RETURN if Zero flag Reset.
RET M	F8	RETURN if Sign flag Set.
RET P	F0	RETURN if Sign flag Reset.
RET PE	E8	RETURN if O/P flag Set.
RET PO	E0	RETURN if O/P flag Reset.

It is an important point in machine code programming to appreciate that a return address that is taken off the stack by a RET instruction, was not necessarily put on the stack by a corresponding CALL instruction.

The following example shows how a table of addresses is handled so that a 'jump' can be made to a required address read from the table.

An example from the 8K monitor program.

In the 'BASIC line scanning routine' the different BASIC commands are allocated to one of seven command classes. Therefore when a certain command is being dealt with, the class number is determined and a 'jump' is made to the appropriate command class routine. The command class numbers are Ø—6, and at ØD16 to ØD1C is a table of displacements values. The following routine is used to access the table, and a RET instruction is used to 'jump' to the required address.

Initially the class number is present in the C register.

address	**Hex.code**	**mnemonic**	**comment**
ØDØ5	21 12 OD	LD HL,+ØD16	Base address.
DØ8	Ø6 ØØ	LD B,+ØØ	Clear B.
DØA	Ø9	ADD HL,BC	Form address.
DØB	4E	LD C,(HL)	Pick-up value.
DØC	Ø9	ADD HL,BC	Form new address.
DØD	E5	PUSH HL	Put in on stack.
DØE	DF	RST ØØ18	Collect next value.
DØF	C9	RET	'Jump' to address.

The Command Class Address Table.

ØD16	17	gives address	ØD2D Class Ø
D17	25	gives address	ØØDC Class 1
D18	53	gives address	ØØDB Class 2
D19	ØF	gives address	ØD28 Class 3
D1A	6B	gives address	ØD85 Class 4
D1B	13	gives address	ØD2E Class 5
D1C	76	gives address	ØD92 Class 6

The BASIC programs that demonstrate these instructions are to be found on page 100.

Group 13: The Rotation Instructions.

In the Z80 instruction set there are many instructions that allow the programmer to rotate the bits in a specified byte.

These instructions can prove to be very useful, especially as rotating a byte to the 'left' has the effect of doubling the value, and rotating a byte to the 'right' has the effect of halving the value.

All of the instructions use and affect the Carry flag in some way, and in some instances it is best considered a 'bit 8' and at other times as a 'bit -1'.

The different types of rotation are best shown diagrammatically and the following diagram shows the direction of the rotation and the position of the Carry flag for each type of rotation.

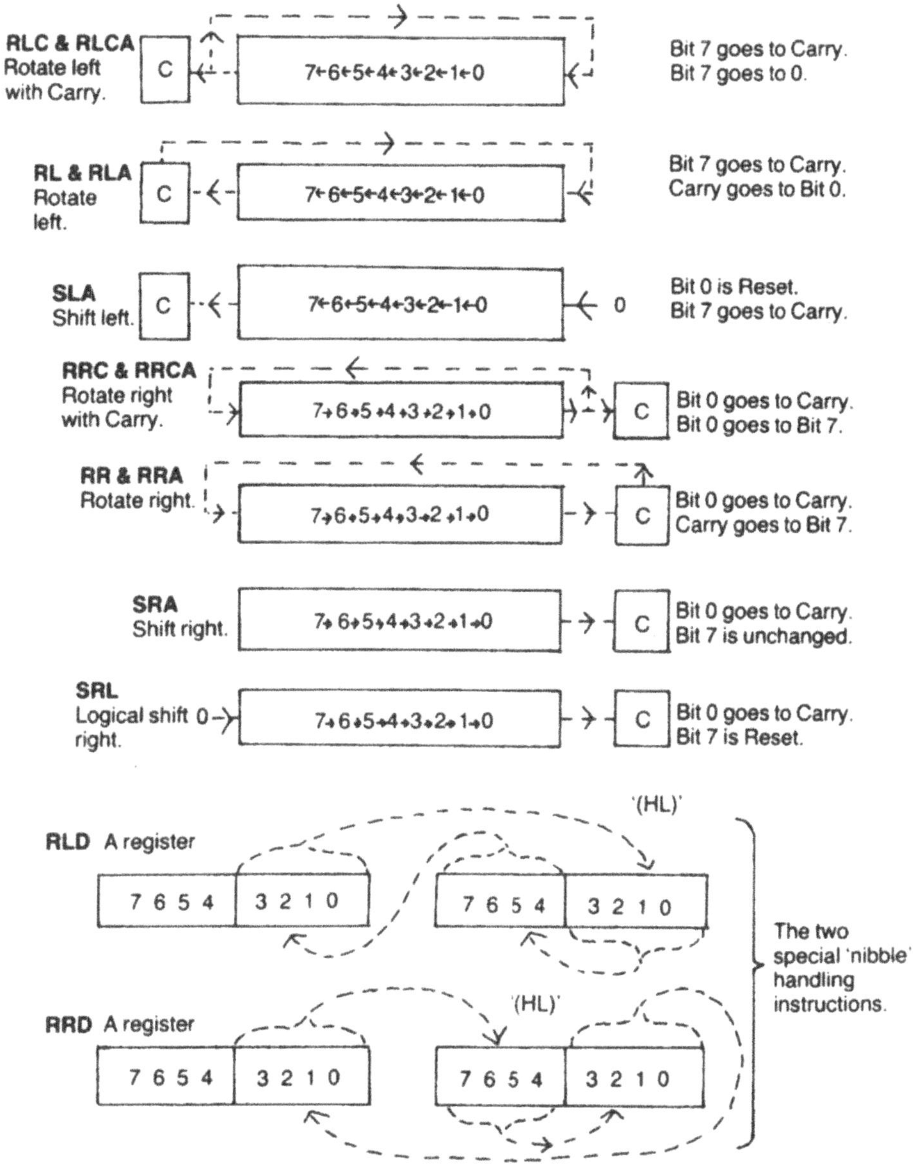

Diagram 9. The Different Types of 'ROTATION'.

The actual instructions are as shown in the table:

Byte	RLC	RL	SLA	RRC	RR	SRA	SRL
A	07	17		OF	1F		
	(RLCA)	(RLA)		(RRCA)	(RRA)		
	or	or		or	or		
	CB 07	CB 17	CB 27	CB OF	CB 1F	CB 2F	CB 3F
H	CB 04	CB 14	CB 24	CB ØC	CB 1C	CB 2C	CB 3C
L	CB Ø5	CB 15	CB 25	CB ØD	CB 1D	CB 2D	CB 3D
B	CB ØØ	CB 1Ø	CB 2Ø	CB Ø8	CB 18	CB 28	CB 38
C	CB Ø1	CB 11	CB 21	CB Ø9	CB 19	CB 29	CB 39
D	CB Ø2	CB 12	CB 22	CB ØA	CB 1A	CB 2A	CB 3A
E	CB Ø3	CB 13	CB 23	CB ØB	CB 1B	CB 2B	CB 3B
(HL)	CB Ø6	CB 16	CB 26	CB ØE	CB 1E	CB 2E	CB 3E
(IX+d)	DD CB	DD CB	DD CB	DD CB	DD CB	DD CB	DD CB
	d Ø6	d 16	d 26	d ØE	d 1E	d 2E	d 3E
(IY+d)	FD CB	FD CB	FD CB	FD CB	FD CB	FD CB	FD CB
	d Ø6	d 16	d 26	d ØE	d 1E	d 2E	d 3E
RRD	ED 67						
RLD	ED 6F						

Note that there are four single byte instructions for rotating the A register.

The following statements can be made about the way the rotation instructions affect the flags.

i. All the instructions, except 'RLD' and 'RRD' affect the Carry flag.

ii. All the two byte instructions affect the Zero flag. The flag being Set if the resultant byte is zero.

iii. All the two byte instructions affect the O/P flag. The flag being Set if the parity of the result is even.

iv. All the two byte instructions affect the Sign flag. The flag copies the sign bit of the resultant byte.

v. The four single byte instructions affect the Carry flag but leave the other flags unchanged.

There are many different ways in which rotation instructions can prove to be useful. The following example shows the 'RL E' instruction being used.

An example from the 8K monitor program.

In the 'SAVE command routine' the program name, and then the program and variables are passed to the cassette output in the following way:

Each byte in turn is loaded into the E register. The Carry flag is Set and the E register rotated left. Bit Ø is thereby Set, and the former bit 7 has been moved to the Carry flag. Bit Ø, at this stage, is a 'marker'. The routine then sends signals to the cassette output that differ depending on whether the Carry flag is Set or Reset.

The Carry flag is then cleared and a jump is made back to the 'RL E' instruction.

Thereby each of the 8 bits of the E register are pushed into the Carry flag in turn. The Zero flag is used to count out the 8 bits as it will be Set when the 'marker' bit is rotated into the Carry flag.

The actual lines are:

address	**Hex.code**	**mnemonic**	**comment**
031E	5E	LD E,(HL)	Pick-up a byte.
31F	37	SCF	SET the Carry flag, 'marker'.
320	CB 13	RL E	Rotate E.
322	C8	RET Z	RETURN after 8 loops.
......			Differing outputs
......			for Carry Set
......			or Reset.
33B	A7	AND A	Clear Carry.
33C	10 FD	DJNZ 033B	A timing delay.
33E	18 E0	JR 0320	Repeat the loop.

(see page 150 for the full 'SAVE' command routine)

A BASIC program that demonstrates the rotation instructions is to be found on page 103.

Group 14: The 'Bit handling' instructions.

The instructions in this group allow the programmer to test for the state of a specified bit, to Reset a specified bit or to Set a specified bit.

Once again the group will be divided into three subgroups.

SUBGROUP A: THE BIT INSTRUCTIONS

These instructions are used to test a specified bit. The result of the test goes to the Zero flag. The flag is SET if the bit tested is RESET, that is holds value zero. The Zero flag is RESET if the bit tested is SET, that is holds value 1.

The actual instruction codes are shown in the following table.

		bit 0	bit 1	bit 2	bit 3	bit 4	bit 5	bit 6	bit 7
A register	RES	87	8F	97	9F	A7	AF	B7	BF
CB **	SET	C7	CF	D7	DF	E7	EF	F7	FF
	BIT	47	4F	57	5F	67	6F	77	7F
H register	RES	84	8C	94	9C	A4	AC	B4	BC
CB**	SET	C4	CC	D4	DC	E4	EC	F4	FC
	BIT	44	4C	54	5C	64	6C	74	7C
L register	RES	85	8D	95	9D	A5	AD	B5	BD
CB**	SET	C5	CD	D5	DD	E5	ED	F5	FD
	BIT	45	4D	55	5D	65	6D	75	7D
B register	RES	80	88	90	98	A0	A8	B0	B8
CB**	SET	C0	C8	D0	D8	E0	E8	F0	F8
	BIT	40	48	50	58	60	68	70	78
C register	RES	81	89	91	99	A1	A9	B1	B9
CB**	SET	C1	C9	D1	D9	E1	E9	F1	F9
	BIT	41	49	51	59	61	69	71	79
D register	RES	82	8A	92	9A	A2	AA	B2	BA
CB**	SET	C2	CA	D2	DA	E2	EA	F2	FA
	BIT	42	4A	52	5A	62	6A	72	7A
E register	RES	83	8B	93	9B	A3	AB	B3	BB
CB**	SET	C3	CB	D3	DB	E3	EB	F3	FB
	BIT	43	4B	53	5B	63	6B	73	7B
(HL)	RES	86	8E	96	9E	A6	AE	B6	BE
CB**	SET	C6	CE	D6	DE	E6	EE	F6	FE
	BIT	46	4E	56	5E	66	6E	76	7E
indexed	RES	86	8E	96	9E	A6	AE	B6	BE
(IX+d)									
DD CB d**	SET	C6	CE	D6	DE	E6	EE	F6	FE
(IY+d)									
FD CB d**	BIT	46	4E	56	5E	66	6E	76	7E

Note that all the instructions are preceded by 'CB', and the instructions for indexed addressed bytes are in addition prefixed by 'DD' or 'FD'.

The use of the BIT instructions is interesting as it enables the programmer to take advantage of the potential offered to him to save RAM and to create rapidly executed programs.

By using a BIT instruction it is possible to use single bits as 'flags'. The following example shows how this is done.

An example from the 8K monitor program.

The system variable 16385 is composed of 8 different flags. Bit 0 of this byte is a flag that is used by the 'PRINT keyword routine' to determine whether or not an extra space is required before a keyword.

e.g. No extra space is required between the line number and the first command in the line 10 IF A THEN GOTO B but an extra space is required between the 'A' and the 'THEN'.

The space between the line number and the first command is always present, but the space between the 'A' and the 'THEN' is an extra space that precedes the keyword.

The following lines show the flag being tested.

address	Hex. code	mnemonic	comment
0951	FD CB 01 40	BIT 0,(4001)	Test the flag.
955	20 02	JR NZ,0959	Jump if needed.
957	AF	XOR A	Clear A.
958	D7	RST 0010	PRINT a space.

959........proceed to print a keyword.

The extra space is printed if the Bit 0 of 16385 is originally RESET, as this leads to the Zero flag being SET and the jump not being made.

SUBGROUP B: THE RES INSTRUCTIONS.

The RES instructions allow the programmer to RESET a specified bit. If the bit however is already in the RESET state then the effect of the execution of a RES instruction will do nothing except reaffirm the situation.

The actual instructions are given in the previous table.

The RES instructions are predominantly used to give a RESET value to a flag.

The following example shows how the 'FAST command routine' uses a RES instruction.

An example from the 8K monitor program.

In the ZX-81 system bits 6 and 7 of system variable 16443, CDFLAG, control the operation of the 'slow' and 'fast' modes.

When the 'FAST command routine' is called, both bit 6 and bit 7 have to be Reset.

The following lines show this being done.

address	Hex. code	mnemonic	comment
0F20	CD E7 02	CALL 02E7	GOSUB 02E7
F23	FD CB 3B B6	RES 6,(403B)	Reset bit 6.
F27	C9	RET	Finished.
02E7	FD CB 3B 7E	BIT 7,(403B)	Test bit 7.
2EB	C8	RET Z	Already 'fast'.

....

2EF	FD CB 3B BE	RES 7,(403B)	Reset bit 7.
2F3	C9	RET	Return.

SUBGROUP C: THE SET INSTRUCTIONS.

The SET instructions allow the programmer to SET a specified bit. If the bit is already SET then the effect of the execution of the SET instruction will do nothing except reaffirm the situation.

The actual instructions are given in the previous table.

The SET instructions are predominantly used to give a SET value to a bit.

The following example shows how the 'SLOW command routine' uses a SET instruction.

An example from the 8K monitor program.

The entry into the 'slow' mode of operation depends on giving a Set value to the bit 6 of the system variable 16443, CDFLAG.

In the 'SLOW command routine' this bit is Set and then a jump is made to the display routine so that a display is produced. If the system was previously in 'fast' mode then the production of a display will be seen to terminate the 'fast' mode. If the system was in 'slow' mode then there will be no detectable change.

The lines are:

address	Hex. code		
0F28	FD CB 3B F6	SET 6,(403B)	Set flag in CDFLAG.
F2C	C3 07 02	JP 0207	Produce a display.

The BASIC program that demonstrates these instructions is to be found on page 104.

Group 15: Block Transferring Instructions and Block Searching Instructions.

The Z80 instruction set contains some very useful instructions that allow the programmer to move blocks of memory or to search blocks of memory.

In order to use the block moving instructions the base address of the block must be in the HL register pair, the address of the destination of the block must be in the DE register pair and the size of the block must be held by the BC register pair.

In order to search a block of memory for the first occurrence of a particular value the base address of the block must be in the HL register pair, the size of the block in the BC register pair and the A register must contain the 'particular value'.

The instructions in the group can be further divided into those that are 'automatic' and those that are 'non-automatic'.

The automatic instructions are so called because a block is moved, or searched, directly the instruction is executed. Therefore only a single instruction is required to move, or search, a block.

The non-automatic instructions only move, or search, one byte for every occasion that the instruction is executed. These instructions therefore require the programmer to create a loop round the instruction if a number of bytes is to be moved, or searched.

The actual instructions are:

Automatic

mnemonic	**instruction** Hex	**comment**
LDIR	ED B0	Block moving – incrementing.
LDDR	ED B8	Block moving – decrementing.
DPIR	ED B1	Block searching – incrementing.
CPDR	ED B9	Block searching – decrementing.

Non-automatic

mnemonic	**instruction** Hex	**comment**
LDI	ED A0	Byte moving – incrementing.
LDD	ED A8	Byte moving – decrementing.
CPI	ED A1	Byte comparing – incrementing.
CPD	ED A9	Byte comparing – decrementing.

Each instruction will now be discussed in turn.

LDIR: This instruction moves a byte from '(HL)' to '(DE)'. The values of HL and DE are then incremented and the counter BC is decremented. When the value of BC reaches zero the moving of bytes stops. The O/P flag will have the Reset value. This instruction can therefore move blocks of data that contain 1-65536 bytes.

LDDR: This instruction is similar to the LDIR instruction except that the values of HL and DE are decremented after each byte is moved.

CPIR: This instruction will automatically search a specified block of memory for the first occurrence of a byte identical to that held in the A register. As each byte is compared the HL register pair is incremented and the BC register pair is decremented.

If a first occurrence is found the search operation stops. The Sign flag is Reset, the Zero flag is Set and the program proceeds with the HL register pair holding the address of the location that follows the location holding the matching byte.

If there are no matching bytes in the whole of the block then the BC register pair will hold zero, the Sign flag is Reset and the O/P flag is Reset. The program then proceeds to the next instruction.

CPDR: This instruction is similar to CPIR except that the HL register pair is decremented before each comparison.

LDI: The execution of this instruction will move a single byte from '(HL)' to '(DE)', the value of the BC register will be decremented, the value of the HL and DE register pairs will be incremented.

If the value of the BC register pair becomes zero then the O/P flag will be Reset, otherwise it will be Set.

LDD: This instruction is similar to LDI except that the HL and DE register pairs are decremented.

CPI: The execution of this instruction will cause the Zero flag to be Set if the byte addressed by the HL register pair matches the contents of the A register. If not the Zero flag is Reset. The HL register pair is incremented and the BC register pair is decremented. The Sign flag is always Reset and if the contents of BC become zero then the O/P flag is Reset.

CPD: This instruction is similar to CPI, except that the HL register pair is decremented.

The use of these instructions is always a little bit complicated, but the instructions are very powerful and are therefore very important.

In the 8K monitor program these instructions are used only occassionally and the following examples are unfortunately not very straightforward.

Examples from the 8K monitor program.

LDDR

Whenever extra space is required in the program area, the variable area or the display file the whole block of data between where the space is needed and the STKEND has to be moved up in the memory.

e.g. when an extra simple variable is added to the end of the variable area, the block of data from the byte holding Hex. 80 to the STKEND has to be moved up 6 locations so as to allow room for the new variable.

The following lines from the 'make space in memory routine' shows the 'LDDR' instruction being used to perform the moving of the block of data. Initially BC holds the size of the space to be added but after the 'change all pointers' routine the BC register pair holds the size of the block.

Before the 'LDDR' is executed the DE register pair will hold the address of the 'new STKEND' and the HL register pair will hold the address of the 'old STKEND'.

address	**Hex. code**	**mnemonic**	**comment**
09A3	CD AD 09	CALL 09AD	'Change all pointers'.
9A6	2A 1C 40	LD HL,(401C)	'New STKEND'.
9A9	EB	EX DE,HL	Exchange values.
9AA	ED B8	LDDR	Move the block.

LDI

The system variable 16419, S-TOP, holds the line number of the top line of a display listing. The location 16419 holding the 'low byte' of the line number and the location 16420 holds the 'high byte'.

However in the program area the line numbers are held with the high number byte before the low number byte.

Therefore when a line number, taken from the program area, is put into system variable 16419 the bytes have to be switched over. This is done using a 'LDI' instruction.

Initially the DE register pair holds the address of the location 16419, Hex. 4023, and the HL register pair points to the 'high byte' of a program line number. Then the 'high byte' is saved temporarily in the A register and the HL register pair incremented to point to the 'low byte' of the program line number. The 'LDI' instruction is then used to move this 'low byte' from '(HL)' to '(DE)'. The execution of this instruction increments the HL and the DE register pairs. As DE now addresses the location 16420 a simple 'LD (DE),A' instruction can be used to move the 'high byte' to its required destination.

The lines involved are:

address	Hex. code	mnemonic	comment
044D	7E	LD A,(HL)	Save 'low byte'.
44E	23	INC HL	Point to 'high byte'.
44F	ED A0	LDI	Move 'high byte'
451	12	LD (DE),A	Move 'low byte'.

CPIR

This instruction is used in the 'D-FILE handling routine' to find the address required by the system variable 16398, DF-CC. This system variable holds the address of the location where the next character is to be placed in the display file.

The following lines show how the program goes back through the display file looking for NEWLINEs and then when it finds the required line, a 'CPIR' instruction is used to go forward along the line.

The contents of the HL register pair is then decremented and saved as DF-CC.

address	Hex. code	mnemonic	comment
0926	04	INC B	The line number required.
927	2B	DEC HL	Go back through the display file, byte by byte.
928	BE	CP (HL)	A holds Hex.76, a NEWLINE.

929	2Ø FC	JR NZ Ø927	Jump back.
92B	1Ø FA	DJNZ Ø927	Each line found.
92D	23	INC HL	Move into line.
92E	ED B1	CPIR	Look for NEWLINE.
93Ø	2B	DEC HL	Back one.
931	22 ØE 4Ø	LD (4ØØE),HL	Save as DF-CC.

The BASIC programs that demonstrate these instructions are to be found on page 105.

Group 16: The Input and Output Instructions.

In the Z80 instruction set there is a comprehensive set of instructions that allow the programmer to collect data from an outside source (IN) or to send data to a peripheral device (OUT).

In the ZX-81 system the incoming data comes from either the keyboard or the cassette player, and the outgoing data goes to the 'logic chip' and hence to the T.V. screen and the cassette player.

There are simple, non-automatic and automatic instructions for both the 'IN' and the 'OUT' types of instruction.

In all cases the data that is moved is handled as a single byte of 8 bits. In the case of an 'IN' instruction the Z80 takes the single byte off the data bus and puts it into a specified register. In the case of an 'OUT' instruction the Z80 puts a copy of the contents of a specified register onto the data bus.

The Z80 shows that it is executing an 'IN' or 'OUT' instruction by using certain of its control signals, and an external device can read these signals and hence prepare to either place a byte of data on the data bus, or to read the byte of data presently on the data bus.

The Z80 also places an address on the address bus whilst it is executing an 'IN' or an 'OUT' instruction. This address is determined by the programmer, and is called the PORT ADDRESS. It is therefore possible for the external devices to read this address and to sense which PORT, or device is to perform the sending or the receiving of the data.

In the following tables of the actual instructions the high and low bytes of the PORT ADDRESS are also shown.

mnemonic	instruction Hex	Input or Output register	High P.A.	Low P.A.
IN A,(+dd)	DB dd	A	A	dd
IN A,(C)	ED 78	A	B	C
IN H,(C)	ED 60	H	B	C
IN L,(C)	ED 68	L	B	C
IN B,(C)	ED 40	B	B	C
IN C,(C)	ED 48	C	B	C
IN D,(C)	ED 50	D	B	C
IN E,(C)	ED 58	E	B	C
INI	ED A2	Non-automatic with increment.		
INIR	ED B2	Automatic with increment.		
IND	ED AA	Non-automatic with decrement.		
INDR	ED BA	Automatic with decrement.		
OUT (+dd),A	D3 dd	A	A	dd
OUT (C),A	ED 79	A	B	C
OUT (C),H	ED 61	H	B	C
OUT (C),L	ED 69	L	B	C
OUT (C),B	ED 41	B	B	C
OUT (C),C	ED 49	C	B	C
OUT (C),D	ED 51	D	B	C
OUT (C),E	ED 59	E	B	C
OUTI	ED A3	Non-automatic with increment.		
OUTIR	ED B3	Automatic with increment.		
OUTD	ED AB	Non-automatic with decrement.		
OUTDR	ED 8B	Automatic with decrement.		

The automatic and non-automatic instructions have been included in the tables but they are not used in the ZX-81 system as their use is primarily involved with the use of discs.

Examples from the 8K monitor program.

'IN'

The 'keyboard scanning routine' gives a very good example of how the 'IN A,(C)' instruction can be used.

In the ZX-81 system the keys of the keyboard are connected to the 8 higher lines of the address bus, and the pressing of the different keys puts different signals on the 5 lower lines of the data bus.

In order to give a different final value for each key with this combination it is necessary to scan the keyboard 8 separate times. On each occasion the values on the 8 higher address lines are changed.

These values are changed using a 'RLC B' instruction, and the counting of the 8 changes is performed by using a 'marker' bit.

Initially bit 0 is the only bit that is Reset, and after 8 rotations it enters the Carry flag and thereby causes an 'exit' from the loop.

The lines involved are:

address	Hex. code	mnemonic	comment
02BE	01 FE FE	LD BC,+FEFE	Initialise.
2C1	ED 78	IN A,(C)	1st Input.
2C3	F6 01	OR +01	
2C5	F6 E0	OR +E0	
....			develop value individual to each key.
....			
2D2	CB 00	RLC B	Rotate.
2D4	ED 78	IN A,(C)	2nd-8th Input.
2D6	38 ED	JR C,02C5	Back until 8th.
....			

(see page 153 for the full 'Keyboard scanning routine')

'OUT'

One of the nice features of using the ZX-81 is that a display of broad bands is produced when a program is being loaded from the cassette player. This 'echoing' of the incoming signal is produced by simply adding an 'OUT (+FF),A' instruction to the 'LOAD command routine'.

The lines are:

address	Hex. code	mnemonic	comment
0350	3E 7F	LD A,+7F	Initialise.
352	DB FE	IN A,(+FE)	Collect byte.
354	D3 FF	OUT (+FF),A	'Echo' the byte.
356................	....		

(see page 151 for the full 'LOAD command routine')

The BASIC programs that demonstrate these instructions are to be found on page 108.

Group 17: The 'Interrupt' instructions.

The Z80 is described as being able to be 'interrupted'. That is to say that the microprocessor can be prevented from proceeding with the sequential execution of the instructions in a particular program by the occurrence of an 'interrupt'.

There are two control lines that go into the Z80 that when 'active' will stop the microprocessor and force it to deal with the interruption.

The first of these lines is called 'NMI' or the 'non-maskable interrupt line'. When this line is activated the Z80 will stop following its present program, it will save the contents of the Program Counter on the stack to be used later as a return address and it will load the Program Counter with the address 0066. The Z80 will always be interrupted when the NMI line is activated and it will always proceed to execute instructions from 0066 onwards.

In the ZX-81 the production of the display during the 'slow' mode uses the NMI facility. When a display is required the NMI line is activated and the 'slow mode display routine' at 0066 is followed. When this routine is finished the 'return address' is collected from the stack and the original program is continued.

The second control line is called 'INT' or the 'maskable interrupt line'. The word 'maskable' implying that the facility can be turned on and off by the programmer. When the power is first connected to the Z80 this interrupt is inactive, or 'disabled', and it requires the execution of a 'EI', enable interrupt, instruction before the facility is active. There is also a 'DI', disable interrupt, that can be used at anytime to run off the interrupt facility. The occurrence of an 'interrupt' will also disable the facility.

There are three programmable modes to the use of the maskable interrupt system. 'Mode O' requires that an external device puts a data byte on the data bus to signify which of the RST addresses is to be put into the Program Counter. 'Mode 1' simply causes the address 0038 to be put into the Program Counter. 'Mode 2' is more complicated and involves a pointer address being formed by the combination of the contents of the I register and the byte placed on the data bus by an external device. This address is then used to index into a vector table and an address held in the table is the address that goes into the Program Counter.

The programmer can select by using the 'IM 0', the 'IM 1' or the 'IM 2' instructions whichever type of interrupt he requires.

The return from an interrupt routine requires a 'RET' instruction to be executed, but there are two special instructions 'RETN' and 'RETI' that may be used if desired. These special instructions have the effect of 'enabling the maskable interrupt' the moment the return has been performed. Therefore if the programmer wishes to 'RETurn from a Non-maskable interrupt' with the 'maskable interrupt' system activated then a 'RETN' instruction should be used. Likewise a 'RETurn from a maskable Interrupt' with reactivation of the interrupt facility requires a 'RETI' instruction rather than a simple 'RET'.

The actual instructions are:

mnemonic	**instruction** Hex.	**Comment**
EI	FB	Enable – maskable interrupts.
DI	F3	Disable – maskable interrupts.
IM 0	ED 46	Mode 0
IM 1	ED 56	Mode 1

IM 2	ED 5E	Mode 2
RETI	ED 4D	Return from maskable interrupt.
RETN	ED 45	Return from non-maskable interrupt.

Examples from the 8K monitor program.

Throughout the monitor program the maskable interrupt facility is used in mode 1. This is specified in the 'initialising routine'.

address	**Hex. code**	**mnemonic**	**comment**
03F6	ED 56	IM 1	Enter mode 1.

In the 'display routine' the maskable interrupt facility is enabled whenever a line of the display is ready to be sent to the T.V. screen. The interrupt always occurs at the end of the display line.

address	Hex. code		comment
0043	FB	EI	Enable interrupt.
44	E9	JP (HL)	Start sending characters.

There are no examples for these instructions in chapter 5.

Group 18: Miscellaneous Instructions.

There are four further instructions in the instruction set.

'CPL'

This instruction allows the programmer to COMPLEMENT the A register.

mnemonic	**instruction** Hex.	**comment**
CPL	2F	Complements A register.

The instruction simply changes all of the bits of the A register to their opposite states. The major flags are not affected.

An example from the 8K monitor program.

In the 'clear memory routine' it is necessary to form the 2's complement of the contents of the BC register pair. This is performed by the following lines:

address	**Hex. code**	**mnemonic**	**comment**
0A61	78	LD A,B	Form
A62	2F	CPL	complement
A63	47	LD B,A	for B.
A64	79	LD A,C	Form
A65	2F	CPL	complement
A66	4F	LD C,A	for C.

A67	03	INC BC	Add the 1 for passing zero.
A68................			

'NEG'

This instruction allows the programmer to form the 2's complement of the A register.

mnemonic	**instruction** Hex.	
NEG	ED 44	2's complement of A

The 'NEG' instruction affects all the flags. The Sign and the Zero flags are affected by the state of the result. The Carry flag will be Set if the A register holds zero before the operation and the O/P flag will be Set if the A register holds Hex.80 before the operation.

'HALT'

The 'HALT' instruction is an interesting instruction as its execution by the Z80 causes the microprocessor to stop executing any further instructions until the occurrence of an interrupt.

mnemonic	**instruction** Hex.
HALT	76

It is no coincidence that the code for NEWLINE in the ZX-81 system is also Hex. 76.

When the characters of a line for the display are being taken out of the display file the circuitry of the ZX-81 makes the Z80 execute 'NOP' instructions. However, when a NEWLINE character is reached, the circuitry allows the Hex. 76 to pass to the Z80 and be executed as a 'HALT' instruction. The Z80 then continues to execute 'NOP' instructions whilst in its 'HALTed' state until it is interrupted. The interruption coming when the R register has counted out the 32 characters for the line.

The 'HALT' instruction is also used in the 'slow mode display routine'. This routine is itself the NMI interrupt routine and the 'HALT' instruction forms part of the timing sequence for the 'slow' mode.

'DAA'

This last instruction is a specialised instruction that allows the programmer to 'Decimally Adjust the A register'.

mnemonic	**instruction** Hex.
DAA	27

In 'binary coded decimal arithmetic' (BCD) the decimal digits 0-9 are represented by the binary 0000 - 1001.

Therefore

the byte 0000 0000 represents decimal 0
the byte 0011 1001 represents decimal 39 and
the byte 1000 1000 represents decimal 88 etc.

A byte containing a 'nibble' (4 bits) of more than 1001 is just not allowed.

The 'Decimal Adjust the A register' simply converts bytes that are in absolute binary arithmetic into BCD.

i.e. if A holds 0000 1010 a 'DAA' operation will give the A register containing: 0001 0000 as this is the BCD representation of decimal 10.

The execution of a 'DAA' instruction does affect the flags. The Sign flag and the Zero flag are simply affected by the result in the usual way. The O/P flag tests the parity of the result. The reaction of the Carry flag is rather complicated as its value is affected by the state of the 'Half carry flag'.

In the 8K monitor program the 'DAA' instruction is only used occasionally and there are no easy examples that can be discussed.

A BASIC program that demonstrates the 'HALT' instruction is to be found on page 108.

5. Demonstration Machine Code Programs

5.1 An outline of the chapter

This chapter contains 26 simple BASIC programs that illustrate the use of many of the machine code instructions of the Z80. All of the programs will RUN on a **1K ZX-81.**

The programs are arranged so that the instructions demonstrated follow the groups described in chapter 4.

All the programs involve POKING machine code instructions into areas of 'free' RAM and then RUNNING the programs by using the USR command.

The machine code instructions are entered as their decimal values, but the Hexadecimal equivalents will always be given together with the mnemonics.

5.2 The Programs

Group 1. The NO OPERATION and RETURN instructions
(see also page 22)

The following simple program shows the RETURN instruction being used as a 'complete' machine code program.

```
Program 1 10 SLOW                         mnemonic     Hex. code
'RET'     20 POKE 17152,201               RET          C9
          30 STOP
          40 LET A=USR 17152
          50 PRINT
          "THE PROGRAM HAS RUN"
```

Enter the above lines and use RUN. Then use RUN 40 to execute the actual machine code program. This program consists of only one machine code instruction – the RET instruction.

The next program shows the use of the NO OPERATION instruction.

```
Program 2 10 SLOW                         mnemonic     Hex. code
'NOP'     20 FOR A=0 TO 98
          30 POKE 17152+A,0               NOP          00
          40 NEXT A
          50 POKE 17251, 201              RET          C9
          60 STOP
          70 LET A=USR 17152
          80 PRINT
          "THE PROGRAM HAS RUN"
```

Enter the above lines and use RUN. Then use RUN 70 to execute the actual machine code program. This time the program is 100 bytes in length, being made up of '99 NO OPERATION instructions and a RETURN'.

These first two programs do not do any useful 'work', but they do show that a machine code program RUN with the USR command requires a final 'RET' instruction in order to make a return to BASIC.

The reader must understand just how the above BASIC programs work before reading further.

Group 2. The instruction for loading registers with constants
(see also page 24)

The USR command returns to the programmer the contents of the BC register pair, as a decimal number. The monitor program however considers the BC register pair to be holding its numbers in absolute binary arithmetic, and this fact must always be remembered.

The first program shows the simple use of a 'LD B,+dd' and a 'LD C,+dd' instruction.

Program 3		mnemonic	Hex. code
'LD	10 SLOW		
C,+dd	20 LET A=17152		
	30 POKE A,14	LD C,+dd	0E
	40 PRINT AT 18,0; "ENTER A VALUE FOR C (0-255)"		
	50 INPUT C		
	60 POKE A+1,C		C
	70 POKE A+2,6	LD B,+dd	06
	80 POKE A+3,0		00
	90 POKE A+4,201	RET	C9
	100 CLS		
	110 PRINT "C NOW CONTAINS"; USR 17152		
	120 RUN		
	RUN		

Note the use of the 'USR 17152' in the PRINT line. This is quite acceptable and can be very helpful.

The second program for this group of instructions shows the 'LD BC,+dddd' instruction.

Program 4		mnemonic	Hex. code
'LD BC,	10 SLOW		
+dddd'	20 LET A=17152		
	30 POKE A,1	LD BC,+dddd	01
	40 PRINT AT 18,0; "ENTER A VALUE FOR BC (0-65535)"		
	50 INPUT BC		
	60 LET B=INT (BC/256)		
	70 LET C=INT (BC-B*256)		
	80 POKE A+1,C		C
	90 POKE A+2,B		B
	100 POKE A+3,201	RET	C9
	110 CLS		
	120 PRINT "BC NOW CONTAINS"; USR 17152		
	130 RUN		
	RUN		

So as to make the working of the above program a little clearer the 'assembler' listing of the 2 instruction line machine code program is given below.

address	Hex. code	mnemonic	comment
4300	01 C B	LD BC,+dddd	Load the constant 'BC'.
303	C9	RET	Return to the BASIC.

Group 3. Register copying and exchanging instructions (see also page 25)

The register copying instructions can be demonstrated by loading registers with constants and then by copying those constants to the B and C registers for returning to the programmer.

The first program shows the 'LD C,L' instruction. In the program a constant is loaded into the L register and then copied into the C register. The B register is loaded with the constant 0 so as to make the program less complicated.

Program 5		mnemonic	Hex. code
'LD C,L'	10 SLOW		
	20 LET A=17152		
	30 POKE A,46	LD L,+dd	2E
	40 PRINT AT 18,0; "ENTER A		
	VALUE FOR L (0-255)"		
	50 INPUT L		
	60 POKE A+1,L		L
	70 POKE A+2,77	LD C,L	4D
	80 POKE A+3,6	LD B,+dd	06
	90 POKE A+4,0		00
	100 POKE A+5,201	RET	C9
	110 CLS		
	120 PRINT "L WAS LOADED		
	WITH"; L		
	130 PRINT		
	140 PRINT "AND NOW		
	C CONTAINS"; USR 17152		
	150 RUN		
	RUN		

The reader is encouraged to use different registers and different instructions in the above program when the principle of the program has been understood.

The second program for this group of instructions demonstrates the important 'EX DE,HL' instruction.

In the program the value entered by the programmer takes a 'tour' round the main registers before emerging unchanged (hopefully).

Program 6		mnemonic	Hex. code
'EX	10 SLOW		
DE,HL'	20 LET A=17152		
	30 POKE A,33	LD HL,+dddd	21
	40 PRINT AT 18,0; "ENTER A		
	VALUE FOR HL (0-65535)"		
	50 INPUT HL		
	60 LET H=INT (HL/256)		
	70 LET L=INT (HL-H*256)		
	80 POKE A+1,L		L
	90 POKE A+2,H		H
	100 POKE A+3,235	EX DE,HL	
	110 POKE A+4,74	LD C,D	4A
	120 POKE A+5,67	LD B,E	43
	130 POKE A+6,81	LD D,C	51
	140 POKE A+7,88	LD E,B	58

```
150 POKE A+8,235                EX DE,HL        EB
160 POKE A+9,68                 LD B,H          44
170 POKE A+10,77                LD C,L          4D
180 POKE A+11,201               RET             C9
190 CLS
200 PRINT "HL WAS
LOADED WITH" ; HL
210 PRINT
220 PRINT "AND NOW
BC CONTAINS" ; USR 17152
230 RUN
RUN
```

The reader is again encouraged to try different instructions in the above program.

Group 4. Instructions for the loading of the registers with data copies from a memory location. (see also page 28).

The instructions in this group allow the programmer to load registers with copies of the contents of locations that are addressed using absolute, indirect or indexed modes of addressing.

The first program demonstrates 'absolute addressing'. In this mode the address of the location is kept as two bytes of data that follow the instruction proper.

Initially a chosen location in memory – address 17152, Hex.4300 – is filled using a POKE command. Than a 'LD A,(addr.)' instruction copies the contents of location 17152 into the A register. Other instructions are then used and the return made to BASIC.

Program 7 'LD A,(addr.)'

```
                                mnemonic        Hex. code
10 SLOW
20 PRINT AT 18,0; "ENTER A
VALUE FOR LOCATION 17152
(0-255)"
30 INPUT VALUE
40 POKE 17152, VALUE
50LET A=17153
60 POKE A,58                    LD A,(addr.)    3A
70 POKE A+1,0                                   00
80 POKE A+2,67                                  43
90 POKE A+3,79                  LD C,A          4F
100 POKE A+4,6                  LD B,+dd        06
110 POKE A+5,0                                  00
120 POKE A,+6,201               RET             C9
130 CLS
```

```
140 PRINT "INPUT VALUE
WAS"; VALUE
150 PRINT
160 PRINT "C REGISTER NOW
CONTAINS"; USR 17153          NOTE: 17153
170 RUN
RUN
```

The second program demonstrates 'indirect addressing'. In this mode the address of the location has to be placed in an 'addressing register pair'.

In the following program the HL register pair is loaded with the address 17152, Hex.4300 and the contents of that location returned to the programmer.

```
Program 8                                      mnemonic       Hex. code
'LD         10 SLOW
C,(HL)'     20 PRINT AT 18,0; "ENTER A
            VALUE FOR LOCATION
            17152 (0-255)"
            30 INPUT VALUE
            40 POKE 17152, VALUE
            50 LET A=17153
            60 POKE A,33                       LD HL,+dddd    21
            70 POKE A+1,0                                     00
            80 POKE A+2,67                                    43
            90 POKE A+3,78                     LD C,(HL)      4E
            100 POKE A+4,6                     LD B,+dd       06
            110 POKE A+5,0                                    00
            120 POKE A+6,201                   RET            C9
            130 CLS
            140 PRINT "INPUT VALUE
            WAS"; VALUE
            150 PRINT
            160 PRINT "C REGISTER NOW
            CONTAINS"; USR 17153
            170 RUN
            RUN
```

The third program demonstrates 'indexed addressing'. In this mode a block of memory is considered to be a table or a list of which the 'base address' is known. The position of the required location must also be known.

In the ZX-81 system, especially in 'slow' mode it is not very practical to change the contents of the IX or the IY register pairs so the following program uses the monitor program's base address of Hex. 4000 and the location that is 'indexed addressed' is in the 'printer buffer', which is at Hex. 403C-405C. (16444-16476).

```
Program 9                                          mnemonic        Hex. code
'LD          10 SLOW
C,(IY+d)'    20 PRINT AT 18,0; "ENTER A
             VALUE FOR LOCATION
             16444 (0-255)"
             30 INPUT VALUE
             40 POKE 16444, VALUE
             50 LET A=17152
             60 POKE A,253                         LD C,(IY+d)     FD
             70 POKE A+1,78                                        4E
             80 POKE A+2,60                                        3C
             90 POKE A+3,6                         LD B,+dd        06
             100 POKE A+4,0                                        00
             110 POKE A+5,201                      RET             00
             120 REM
             130 CLS
             140 PRINT "INPUT VALUE
             WAS"; VALUE
             150 PRINT
             160 PRINT "ENTRY D= " "3C" "
             IS";USR 17152                         (shifted Qs.)
             170 RUN
             RUN
```

The reader is encouraged to change the parameters in lines 20, 40 80 and 160, to examine other locations.

Group 5. Instructions for loading locations in memory with data copied from registers, or with constants. (see also page 33)

The instructions in this group allow the programmer to load memory locations with data copied from registers and to load addressed locations with constants.

Once again absolute, indirect and indexed addressing can be used.

The first program shows the 'LD (addr.),A' instruction. In the program the user is asked to enter a 'VALUE' and an 'ADDRESS' and then the machine code routine will 'load' the specified location. A PEEK command is used to show the user that the operation is successful.

Program		mnemonic	Hex. code
10	10 SLOW		
LD	20 PRINT AT 18,0; ENTER A		
(addr.),A'	VALUE (0-255)"		
	30 INPUT VALUE		
	40 CLS		
	50 LET A=17152		
	60 POKE A,62	LD A,+dd	3E
	70 POKE A+1, VALUE		VALUE
	80 PRINT AT 18,0; "ENTER THE ADDRESS OF A LOCATION (16384-17407)"		
	90 INPUT ADDRESS		
	100 CLS		
	110 LET H=INT (ADDRESS/256)		
	120 LET L=INT (ADDRESS-H*256)		
	130 POKE A+2,50	LD (addr.),A	32
	140 POKE A+3,L		L
	150 POKE A+4,H		H
	160 POKE A+5,201	RET	C9
	170 LET K=USR 17152		
	180 PRINT "INPUT VALUE WAS"; VALUE		
	190 PRINT		
	200 PRINT "LOCATION"; ADDRESS; "HOLDS"; PEEK ADDRESS		
	210 RUN		
	RUN		

The user of the above program must be careful to choose a value for 'ADDRESS' that is sensible. In this particular case locations 17158 to 17300 are definitely 'free' locations.

The second program shows 'indirect addressing' being used to address the required location. A 'LD (HL),E' instruction is used to 'load' that location.

Program		mnemonic	Hex. code
11	10 SLOW		
'LD	20 PRINT AT 18,0; "ENTER A		
(HL),E'	VALUE (0-255)"		
	30 INPUT VALUE		
	40 CLS		

```
50 LET A=17152
60 POKE A,30                      LD E,+dd        1E
70 POKE A+1, VALUE                                VALUE
80 PRINT AT 18,0; "ENTER
THE ADDRESS OF A
LOCATION (16384-17407)"
90 INPUT ADDRESS
100 CLS
110 LET H=INT
(ADDRESS/256)
120 LET L=INT
(ADDRESS-H*256)
130 POKE A+2,33                   LD HL,+dddd     21
140 POKE A+3,L                                    L
150 POKE A+4,H                                    H
160 POKE A+5,115                  LD (HL),E       73
170 POKE A+6,201                  RET             C9
180 LET K=USR 17152
190 PRINT "INPUT VALUE
WAS"; VALUE
200 PRINT
210 PRINT "LOCATION",
ADDRESS;" HOLDS";
PEEK ADDRESS
220 RUN
RUN
```

Once again the user must choose the value for 'ADDRESS' sensibly. Locations 17159 to 17300 are 'free' locations.

The third program for this group uses 'indexed addressing and demonstrates the 'LD (IY+d),+dd' instruction. Again the 'base address' of Hex. 4000 is used and the indexed locations all form part of the 'printer buffer'.

```
Program                                     mnemonic     Hex. code
12          10 SLOW
            20 PRINT "ENTRY NO.",
 LD         "CONTENTS"
(IY+d),     30 PRINT
+dd'        40 FOR I=60 TO 70
            50 PRINT I, PEEK (16384+I)
            60 NEXT I
            70 PRINT AT 18,0; "ENTER A
            VALUE (0-255)"
            80 INPUT VALUE
```

```
90 PRINT AT 18,0; "WHICH
LOCATION? (60-70)"
100 INPUT D
110 IF D<60 OR D>70 THEN
GOTO 100
120 LET A=17252 NOTE: 17252
130 POKE A,253                LD(IY+d),
                              +dd          FD
140 POKE A+1,54                            36
150 POKE A+2,D                             D
160 POKE A+3, VALUE                        VALUE
170 POKE A+4,201              RET          C9
180 LET K=USR 17252
190 CLS
200 RUN
RUN
```

In the above program the user can 'load' values into a part of the printer buffer'. Note that in the program the 'displacement' value D is in decimal arithmetic. The locations would more normally be described in Hex. as 'IY+3C' to 'IY+46;.

Group 6: The addition instructions.
(see also page 36)

There are three subgroups of instructions within this group. The ADD instructions perform straightforward addition operations in absolute binary arithmetic. The INC instructions simply increment the byte or bytes specified. The ADC perform addition together with incrementation if the Carry flag is Set.

The first BASIC program shows the 'ADD A,+dd' instruction.

```
Program                                     mnemonic      Hex. code
13        10 SLOW
 ADD      20 PRINT AT 18,0; "ENTER
A,+dd'    FIRST VALUE (0-255)"
          30 INPUT F
          40 PRINT AT 18,0; "ENTER
          SECOND VALUE (0-255)"
          50 INPUT S
          60 CLS
          70 LET A=17152
          80 POKE A,62                      LD A,+dd      3E
          90 POKE A+1,F                                   F
          100 POKE A+2,198                  ADD A,+dd     C6
          110 POKE A+3,S                                  S
```

```
120 POKE A+4,79                  LD C,A        4F
130 POKE A+5,6                   LD B,+dd      06
140 POKE A+6,0                                 00
150 POKE A+7,201                 RET           C9
160 PRINT "ADDITION";
F;" + "; S;"="; USR 17152
170 RUN
RUN
```

In the above program the first value, F, is loaded into the A register. The second value becomes the data byte associated with the 'ADD A,+dd' instruction. The program does show the 'addition in absolute binary arithmetic' quite truthfully.

The second program for this group demonstrates the 'INC BC' instruction. Again the operation is performed in absolute binary arithmetic.

```
Program                                    mnemonic      Hex. code
14         10 SLOW
'INC BC'   20 PRINT AT 18,0; "ENTER
           VALUE FOR BC (0-65535)"
           30 INPUT BC
           40 CLS
           50 LET B=INT (BC/256)
           60 LET C=INT (BC-B*256)
           70 LET A=17152
           80 POKE A,1                      LD BC,+dddd   01
           90 POKE A+1,C                                  C
           100 POKE A+2,B                                 B
           110 POKE A+3,3                   INC BC        03
           120 POKE A+4,201                 RET           C
           130 PRINT "OLD CONTENTS
           OF BC = "; BC
           140 PRINT
           150 PRINT "WHEN
           INCREMENTED BC=";
           USR 17152
           160 RUN
           RUN
```

Try entering a value of 65535.

There is no program to illustrate the 'ADC' instruction but an 'ADC +dd' instruction is used in the first program for the next group of instructions.

Group 7: The subtraction instructions.
(see also page 40)

Once again there are three subgroups of instruction in this group.

The SUB instructions perform straightforward subtraction. The DEC instructions simply decrement the specified byte or bytes. The SBC instructions perform subtraction and also decrementation if the Carry flag is Set.

The use of the Carry flag is very important and the first program shows how:

operations that are
F > S give Carry Reset.
F = S gives Carry Reset and
F < S give Carry Set.

(F is a first value, S is a second value)

All subtraction operations are in 'absolute binary arithemtic'.

In the first program, that illustrates the 'SUB +dd' instruction, there is a change in the method of loading the machine code into the ZX-81.

Line 10, the first line in the BASIC program is a REM statement that has 16 characters that are used to reserve space for the machine code program.

The change is required as in a standard IK ZX-81 there is very little 'free RAM' available when a large BASIC is entered, and it is a good idea to reserve space by using a 'dummy' REM line.

Program		mnemonic	Hex. code
15	10 REM 1234567890123456	(16 bytes)	
SUB	20 SLOW		
+dd'	30 PRINT AT 18,0; "ENTER VALUE ONE (0-255)"		
	40 INPUT F		
	50 PRINT AT 18,12;"TWO"		
	60 INPUT S		
	70 CLS		
	80 LET A=16515 Note: 16515		
	90 POKE A,62	LD A,+dd	3E
	100 POKE A+1,F		F
	110 POKE A+2,214	SUB +dd	D6
	120 POKE A+3,S		S
	130 POKE A+4,79	LD C,A	4F
	140 POKE A+5,6	LD B,+dd	06
	150 POKE A+6,0		00
	160 POKE A+7,62	LD A,+dd	3E
	170 POKE A+8,0		00
	180 POKE A+9,206	ADC A,+dd	CE
	190 POKE A+10,0		00

```
200 POKE A+11,50          LD (addr.),A      32
210 POKE A+12,130
                          (1st character)   82
220 POKE A+13,64          (in line 10)      40
230 POKE A+14,201         RET               C9
240 PRINT "SUBTRACTION";
F;"-";S;"=";USR 16515,,
"CARRY"; Note: USR 16515
250 IF NOT PEEK 16514
THEN PRINT "RE"
260 PRINT "SET"
270 RUN
RUN
```

Try the above program with;
24 — 12 to give 12 & Carry Reset
24 — 24 to give 0 & Carry Reset and
24 — 25 to give 255 & Carry SET

In the program F takes the value of the First value entered. S takes the value of the Second value.

A 'SUB +dd' instruction performs the subtraction. The result is then put into the BC register pair.

The value of the Carry flag is then found by using a 'ADC+00' instruction, this value is stored in location 16514 and subsequently read with a PEEK command.

The DEC instructions are so similar to the INC instructions (though opposite in action) that the following changes to program 14 can be used if wished.

```
                          mnemonic          Hex. code
110 POKE A+3,11           DEC BC            0B
150 PRINT "WHEN
DECREMENTED BC=";
USR 17152
```

There is no program to demonstrate the SBC instructions as they are commonly used as straighforward subtraction instructions by using 'AND A' before the 'SBC' so as to clear the Carry flag.

Group 8; The Compare instructions. (see also page 44)

The Compare instructions in effect perform a simple subtraction operation, set the state of the flags and then discard the result.

As the compare operation is so similar to subtraction the instructions can be demonstrated by modifying program 15.

Therefore change in program 15:

```
                                        mnemonic     Hex. code
110 POKE A+2,254                        CP +dd       FE
240 PRINT "COMPARISON",
F;":"; S,,"CARRY";
245 LET K=USR 16515
```

Again try 'greater than', 'equals' and 'less than' comparisons and see how the state of the Carry flag changes.

Group 9; The Logical instructions. (see also page 45)

The instructions in this group allow the programmer to perform the logical operations of AND, OR and XOR.

The first program shows the 'AND +dd' instruction. In this BASIC program the 'free RAM' of the 'printer buffer' is used to hold the machine code program.

Program 16 'AND +dd'

```
                                        mnemonic     Hex.code
 10 SLOW
 20 PRINT AT 18,0; "ENTER
    VALUE ONE (0—255)"
 30 INPUT F
 40 PRINT AT 18,12; "TWO"
 50 INPUT S
 60 CLS
 70 LET A=16445
 80 POKE A,62                           LD A,+dd     3E
 90 POKE A+1,F                                       F
100 POKE A+2,230                        AND +dd      E6
110 POKE A+3,S                                       S
120 POKE A+4,50                         LD (addr.),A 32
130 POKE A+5,60 (1st location                        3C
                of PRBUFF)
140 POKE A+6,64                                      40
150 POKE A+7, 201                       RET          C9
160 LET K=16445
170 PRINT "LOGICAL";F;
    "AND";S;"IS";PEEK 16444
180 RUN
```

The above program works in decimal arithmetic, but the logical operations are performed in binary by the Arithmetic-logic-unit of the Z80.

To demonstrate the OR operation the following changes can be made to program 16.

```
100  POKE A+2,246                  OR +dd         F6
170  PRINT"LOGICAL";F;"OR"
     ;S;"IS";PEEK 16444
```

To demonstrate the XOR operation

```
100  POKE A+2,238
170  PRINT "LOGICAL";F;
"XOR";S;"IS";PEEK 16444
```

Group 10; The Jump instructions. (see also page 49)
In this group are the unconditional and the conditional Jump instructions, and the jumps made can be to an absolute addressed location or to a 'relative' addressed location.

The following program shows how the different results of a simple addition operation set the flags and hence permit or otherwise a jump to be made.

The program initially asks the user to specify an instruction, and the following are permitted instructions for this program:

decimal	mnemonic	Hex.code
24	JR e	18 e
32	JR NZ,e	20 e
40	JR Z,e	28 e
48	JR NC,e	30 e
56	JR C,e	38 e

The decimal value must be entered.

Program 17
'JR e'

```
                                         mnemonic       Hex.code
10  SLOW
20  PRINT AT 18,0; "ENTER
INSTRUCTION","E.G.
     <40> FOR <JR Z,E>"
30  INPUT B
40  CLS
50  PRINT AT 18,0; ENTER
VALUE ONE (0-255)"
60  INPUT F
70  PRINT AT 18,12; "TWO"
80  INPUT S
90  CLS
100  LET A=16445
110  POKE A,1                            LD BC,+dddd    01
120  POKE A+1,0                                         00
130  POKE A+2,0                                         00
140  POKE A+3,62                         LD A,+dd       3E
```

```
150 POKE A+4,F                                  F
160 POKE A+5,198            ADD A,+dd           C6
170 POKE A+6,S                                  S
180 POKE A+7,50             LD (addr.),A        32
190 POKE A+8,60                                 3C
200 POKE A+9,64                                 40
210 POKE A+10,B             JR e                B
220 POKE A+11,1                                 1
230 POKE A+12,201           RET                 C9
240 POKE A+13,3             INC BC              03
250 POKE A+14,201           RET                 C9
260 LET C=USR 16444
270 PRINT F;"+";S;"="PEEK 16444
280 IF NOT C THEN PRINT
"NO";
290 PRINT "JUMP"
300 RUN
RUN
```

Note: The user must be careful to enter only valid instructions. SAVEing the program is advised.

In the above program the BC register pair is initially set to contain zero. However if a jump is made the value in BC is incremented, and the incremented value returned to the user.

The above program can be simply changed to give examples of the 'absolute' addressing instructions.

The following instructions are then valid:

decimal	**mnemonic**	**Hex. code**
195	JP addr.	C3 addr.
194	JP NZ, addr.	C2 addr.
202	JP Z, addr.	CA addr.
210	JP NC, addr.	D2 addr.
218	JP C, addr.	DA addr.
226	JP PO, addr.	E2 addr.
234	JP PE, addr.	EA addr.
242	JP P, addr.	F2 addr.
250	JP M, addr.	FA addr.

The changes to program 17 are:

```
20 PRINT AT 18,0;"ENTER
INSTRUCTION", "E.G. <250>
FOR <JP M,ADDR.> "
```

```
220 POKE A+11,75 (location A+14)
                                        4B
230 POKE A+12,64                        40
240 POKE A+13,201      RET              C9
250 POKE A+14,3        INC BC           03
255 POKE A+15,201      RET              C9
```

The user can alter the location A+5 if wished.

```
E.g. changes lines:
160 POKE A+5,214
270 PRINT F;" - ";S;" = ";PEEK
16444
```

will change the addition operation to subtraction.

Group 11; The 'DJNZ e' instruction (see also page 59)
This very commonly used instruction can be considered to be equivalent to the BASIC NEXT command.

The following very simple program shows a FOR . . . NEXT loop being used to SUM a series.

```
10 SLOW
 20 PRINT "THE SUM OF THE
SERIES","1,2 . . . 254,255"
 30 PRINT
 40 PRINT "=";
 50 INPUT A$
 60 LET SUM=0
 70 FOR C=255 TO 1 STEP
-1
 80 LET SUM=SUM+C
 90 NEXT C
100 PRINT SUM
```

The program includes the line 50 INPUT A$ as a waiting point for the user — simply press NEWLINE.

As written the second half of the program takes 9 seconds to produce the answer 32640.

The following program demonstrates the 'DJNZ e' instruction being used in the equivalent machine code program.

```
            10 SLOW
            20 PRINT "THE SUM OF
           THE SERIES","1,2 . . . 254,255"
            30 PRINT
Program     40 PRINT "=";
18          50 LET A=16444
'DJNZ e'    60 POKE A,33              LD HL,+dddd   21
```

```
 70 POKE A+1,0                        00
 80 POKE A+2,0                        00
 90 POKE A+3,85         LD D,L        55
100 POKE A+4,6          LD B,+dd      06
110 POKE A+5,255                      FF
120 POKE A+6,88         LD E,B        58
130 POKE A+7,25         ADD HL,DE     19
140 POKE A+8,16         DJNZ e        10
150 POKE A+9,252                      FC
160 POKE A+10,68        LD B,H        44
170 POKE A+11,77        LD C,L        4D
180 POKE A+12,201       RET           C9
190 INPUT A$
200 PRINT USR 16444
```

As written the above program takes in 'slow' mode about a second. However by changing line 10 to; 10 FAST and running the program again, it will be seen to be very much quicker.

The above program is worth studying in further detail as it really is the first functional program in the book.

The 'assembler' listing is as follows;

address	Hex. code	Label	mnemonic	Comment
403C	21 00 00	START	LD HL,,+0000	Clear HL for answer.
03F	55		LD D,L	Set D to zero.
040	06 FF		LD B,+FF	For 255 times.
042	58	FOR	LD E,B	Transfer to low register.
043	19		ADD HL,DE	For part answers.
044	10 FC	NEXT	DJNZ FOR	Go back.
046	44		LD B,H	Transfer to USR.
047	4D		LD C,L	Transfer to USR.
048	C9	END	RET	Return.

This 13 byte routine can be considered to be a subroutine to be used on any occasion where there is the need to sum a series of suitable numbers.

Group 12; The Stack instructions. (see also page 61)

The first subgroup of instructions in this group consist of those instructions that can be used by the programmer to store data temporarily on the stack. The second subgroup consists of those instructions that use the stack themselves to hold 'return addresses'.

The following program shows the stack being used to hold the value entered by the user, this value is then 'popped' off the stack and returned to the user.

The program shows how a value can be moved from one register pair to another by the use of the appropriate 'PUSH' and 'POP' instructions.

Program 19 'PUSH HL'

```
                                            mnemonic      Hex. code
 10 SLOW
 20 PRINT AT 18,0; "ENTER A
VALUE FOR HL (0-65535)"
 30 INPUT HL
 40 CLS
 50 LET H=INT (HL/256)
 60 LET L=INT (HL=H*256)
 70 PRINT "HL WAS FILLED
    WITH";HL
 80 PRINT
 90 PRINT "BC NOW
    CONTAINS";
100 LET A=16444
110 POKE A,33                               LD HL,+dddd   21
120 POKE A+1,L                                            L
130 POKE A+2,H                                            H
140 POKE A+3,229                            PUSH HL       E5
150 POKE A+4,193                            POP HL        C1
160 POKE A+5,201                            RET           C9
170 PRINT USR 16444
180 RUN
```

The second program demonstrates the absolute CALL and the conditional CALL instructions. The program is similar to program 17 that showed the conditional JP instructions.

The user is first asked to enter the 'instruction', in decimal, and then the values for an addition operation. The program then shows whether or not the CALL instruction would be used to run a subroutine.

The following instructions are valid for the program:

decimal	**mnemonic**	**Hex**
205	CALL addr.	CD addr.
220	CALL C, addr.	DC addr.
212	CALL NC, addr.	D4 addr.
204	CALL Z, addr.	CC addr.
196	CALL NZ, addr.	C4 addr.
236	CALL PE, addr.	EC addr.
228	CALL PO, addr.	E4 addr.
252	CALL M, addr.	FC addr.
244	CALL P, addr.	F4 addr.

The user is well advised to SAVE the program before running it for the first time.

If the subroutine is **not** called the program will print; 'NO CALL'.

However whenever the subroutine has been called the program will print the simple addition operation.

Program 20 'CALL addr.'

```
                                          mnemonic        Hex. code
 10  SLOW
 20  PRINT AT 18,0; "ENTER
    INSTRUCTION", "E.G. <252>
      FOR <CALL M,ADDR.>"
 30  INPUT B
 40  CLS
 50  PRINT AT 18,0;"ENTER
     VALUE ONE (0-255)"
 60  INPUT F
 70  PRINT AT 18,12;"TWO"
 80  INPUT S
 90  CLS
100  LET A=16445
110  POKE A,1                         LD BC+dddd       01
120  POKE A+1,0                                        00
130  POKE A+2,0                                        00
140  POKE A+3,62                      LD A,+dd         3E
150  POKE A+4,F                                        F
160  POKE A+5,198                     ADD A,+dd        C6
170  POKE A+6,S                                        S
180  POKE A+7,B                       CALL xxxxx       B
190  POKE A+8,72 (address of)                          48
200  POKE A+9,64 (A+11)                                40
210  POKE A+10,201                    RET              C9
220  POKE A+11,3                      INC BC           03
230  POKE A+12,50                     LD (addr.),A     32
240  POKE A+13,60                     3C
                                (address of)
250  POKE A+14,64               (A-1)                  40
260  POKE A+15,201                    RET              C9
270  LET C=USR 16445
280  IF C THEN PRINT F;" +
     ";S;"=";PEEK 1644
290  IF NOT C THEN PRINT
     "NO CALL"
300  RUN
RUN
```

In the program the result of the addition is stored in location 16444, from where it is read using a PEEK command if the subroutine was called. The value returned to the user in BC is used as a flag to show whether or not the subroutine was called.

It is not possible to demonstrate the RST instructions as they all are 'preprogrammed' in the 8K ROM.

The exercise of writing a BASIC program to demonstrate the conditional RET instructions is left to the reader.

Hint: Try changing program 20. Location A+7 could have a 'CALL addr.' instruction and location A+10 could have a conditional RET instruction.

Group 13; The Rotation instructions. (see also page 66)

There are seven main types of rotation. The following demonstration program shows the results of rotations on the C register.

The user is allowed to Reset, or Set, the Carry flag prior to the rotation.

As before the user is initially asked to enter the instruction to be used. The following instructions are valid

decimal	**mnemonic**	**Hex. code**
1	RLC C	CB Ø1
17	RL C	CB 11
33	SLA C	CB 21
9	RRC C	CB Ø9
25	RR C	CB 19
41	SRA C	CB 29
57	SRL C	CB 39

Program 21 'RLC C'

```
 1Ø SLOW
 2Ø PRINT AT 18,Ø;
    "ENTER
    INSTRUCTION",
    "E.G<1> FOR<RLC C">
 3Ø INPUT B
 4Ø CLS
 5Ø PRINT AT 18,Ø;
    "ENTER VALUE FOR
    CARRY (Ø OR 1)"
 6Ø INPUT C
 7Ø CLS
 8Ø PRINT AT 18,Ø;
    "ENTER VALUE FOR
    C (Ø-255)"
 9Ø INPUT D
1ØØ CLS
```

```
110 LET A=16445
120 POKE A,6                 LD B,+dd        06
130 POKE A+1,0                               00
140 POKE A+2,167-112*C       XOR A or SCF    A7 or 37
150 POKE A+3,14              LD C            0E
160 POKE A+4,D                               D
170 POKE A+5,203             Rotation        CB
180 POKE A+6,B                               B
190 POKE A+7,62              LD A,+dd        3E
200 POKE A+8,0                               00
210 POKE A+9,206             ADC +dd         CE
220 POKE A+10,0                              00
230 POKE A+11,50             LD (addr.),A    32
240 POKE A+12,60 (location of)
                                             3C
250 POKE A+13,64 (A-1)                       40
260 POKE A+14,201            RET             C9
270 PRINT D;"ROTATES
    TO";USR
    16445;"CARRY";
280 LET E=PEEK 16444
290 IF NOT E THEN PRINT
    "SET"
300 IF E THEN PRINT
    "RESET"
310 RUN
RUN
```

In the program the result of the rotation is stored in location 16444. The state of the Carry flag is read using an 'ADC +dd' instruction as before.

The user is again advised to SAVE the program before using it for the first time, as the basic1K program does not have any 'error checking' on the input values.

The user might like to try: Instruction=1, Carry=0, C=128 & Instruction=57, Carry=1, C=1 etc.

Group 14; the 'Bit handling' instructions. *(see also page 69)*

This group contains the BIT, RES and SET instructions. The BIT instructions allow the programmer to 'test' a particular bit in a specified byte. The RES and SET instructions allow the programmer to Reset or Set a particular bit.

The following demonstration program shows the BIT type of instruction being used to convert a decimal number to its binary equivalent.

```
                                          mnemonic       Hex. code
          10  SLOW
          20  PRINT AT 16,0;"DECIMAL
              TO BINARY
              CONVERSION"
          30  PRINT AT 18,0;"ENTER
              DECIMAL NUMBER
              (0-255)"
          40  INPUT B
Program   50  CLS
22        60  PRINT AT 6,6;B;"   "; (3 spaces)
'BIT'     70  LET A=16444
          80  POKE A,1                LD BC,+dddd    01
          90  POKE A+1,0                             00
         100  POKE A+2,0                             00
         110  POKE A+3,62             LD A,+dd       3E
         120  POKE A+4,B                             B
         130  POKE A+5,203            BIT xxxx       CB
         140  POKE A+7,200            RET Z          C8
         150  POKE A+8,3              INC BC         03
         160  POKE A+9,201            RET            C9
         170  FOR C=1 TO 8
         180    POKE A+6,135-C*8   (the eight BIT instructions)
         190  PRINT USR 16444;
         200  NEXT C
         210  RUN

              RUN
```

In the program each bit of the byte containing B is tested in turn. The value of the BC register pair is then incremented if the bit is Set. The above program runs very slowly in 'slow' mode and would be very much faster if written totally in machine code.

There are no demonstration programs to show the RES and SET instructions, but the reader is encouraged to write his own.

Group 15; Block Transferring instructions and Block Searching instructions (see also page 72)

The instructions in this group are very important. They are however rather difficult instructions to use, unless the programmer has a clear understanding of just what he is trying to do.

The first program shows the 'LDIR' instruction being used to move a block of code within the BASIC program area.

Enter these lines:

```
10 REM 12345678901234567890123
20 REM ***COPY THIS MESSAGE***
30 SLOW
40 FOR I=16514 TO 16536
50 PRINT CHR$ PEEK I;
60 NEXT I
70 PRINT
80 FOR I=16543 TO 16565
90 PRINT CHR$ PEEK I;
100 NEXT I
RUN
```

When the above program is RUN the display will show the characters from lines 10 and 20.

The addresses of the locations for the 'start' of these lines are therefore; Line 10, decimal 16514, Hex. 4082, Line 20, decimal 16543, Hex. 409F and the length of the lines are; decimal 23, Hex.0017.

With this information known enter the following demonstration program which will move the contents of line 20 to line 10 using a 'LDIR' instruction.

```
                                                    mnemonic       Hex. code
           10 REM 12345678901234567890123
           20 REM ***COPY THIS MESSAGE***
Program    30 SLOW
23         40 LET A=16444
'LDIR'     50 POKE A,33                             LD HL,+dddd    21
           60 POKE A+1,159 (start of)                              9F
           70 POKE A+2, 64 (line 20)                               40
           80 POKE A+3,17                           LD DE,+dddd    11
           90 POKE A+4,130 (start of)                              82
          100 POKE A+5,64 (line 10                                 40
          110 POKE A+6,1                            LD BC,+dddd    01
          120 POKE A+7,23                                          17
          130 POKE A+8,0                                           00
          140 POKE A+9,237                          LDIR           ED
          150 POKE A+10,176                                        B0
          160 POKE A+11,201                         RET            C9
          170 LET K=USR 16444
          180 LIST
          RUN
```

When the program is RUN it will be seen that the line 20 has been copied into the 23 reserved locations of line 10.

The 'assembler' listing is given below.

address	**Hex. code**	**mnemonic**	**comment**
403C	21 9F 40	LD HL,409F	Start of line 20.
03F	11 82 40	LD DE,4082	Start of line 10.
042	01 17 00	LD BC,0017	The number of characters.
045	ED BO	LDIR	Block move.
047	C9	RET	Return to BASIC.

The reader is encouraged to try his own programs using 'LDIR', 'LDDR', 'LDI' and 'LDD'.

The second program shows the use of the 'CPIR' instruction.

In the program a search is made in the 8K ROM for the first occurrences of the numbers 0-255.

		mnemonic	Hex. code
	10 SLOW		
	20 LET A=16444		
	30 POKE A,62	LD A,+dd	3E
Program	40 POKE A+2,1	LD BC,+dddd	01
24	50 POKE A+3,0 (the 8K)		00
'CPIR'	60 POKE A+4,32 (ROM)		32
	70 POKE A+5,33	LD HL,+dddd	21
	80 POKE A+6,0		00
	90 POKE A+7,0		00
	100 POKE A+8,237	CPIR	ED
	110 POKE A+9,177		B1
	120 POKE A+10,68	LD B,H	44
	130 POKE A+11,77	LD C,L	4D
	140 POKE A+12,201	RET	C9
	150 FOR B=0 TO 255		
	160 POKE A+1,B (match to this byte)		
	170 LET C=USR 16444		
	180 IF NOT C=8192 THEN PRINT B;"OCCURS FIRST AT";C−1		
	190 IF C=8192 THEN PRINT B;"DOES NOT OCCUR"		
	200 PAUSE 100		
	210 NEXT B		

Note that in line 18Ø that the location that holds the matching is given by 'C-1', as the HL register pair is incremented before the comparison is made.

The above program is purposely simple and does not check the flags in the case of there being 'no match'.

It is of interest that the value decimal 153, Hex. 99, does just not occur in the whole of the monitor program.

Group 16; The Input and Output instructions. (see also page 76)

The instructions in this group allow the programmer to accept bytes of data from an external source, and to send bytes out from the Z80.

The following simple program uses the 'OUT (+FF),A' instruction to produce a 'hum' that can be recorded on a cassette tape.

The program just starts to show how 'music' can be produced.

		mnemonic	Hex. code
	1Ø FAST Note; FAST		
	2Ø LET A=16444		
Program	3Ø POKE A,6	LD B,+dd	Ø6
25	4Ø POKE A+1,255		FF
'OUT	5Ø POKE A+2,211	OUT (+FF),A	D3
(+FF),A	6Ø POKE A+3,255		FF
	7Ø POKE A+4,16	DJNZ e	1Ø
	8Ø POKE A+5,252		FC
	9Ø POKE A+6,2Ø1	RET	C9
	1ØØ FOR B=1 TO 3ØØ		
	11Ø LET K=USR 16444		
	12Ø NEXT B		
	RUN		

Group 17; The 'Interrupt' instructions. (see also page 78)

There are no demonstration programs for these as there is no facility for the user to use the interrupt lines in a standard 1K ZX-81.

Group 18; Miscellaneous instructions. (see also page 80)

The only demonstration program for this group is the following program that shows the 'HALT' instruction being used in the 'slow' mode.

In 'slow' mode the NMI is used to split the machine code into blocks. In the following program a 'HALT' instruction is used, however if the program is run in 'fast' mode the program will be lost, showing that there are no interrupts in 'fast' mode.

Program 26 'HALT'

```
 10 SLOW
 20 LET B=3
 30 LET A=16444
 40 POKE A,118            HALT     76
 50 POKE A+1,201          RET      C9
 60 LET K=USR 16444
 70 CLS
 80 PRINT AT
    6-B,5;"INTERRUPT
    WORKING"
 90 LET B=-B
100 GOTO 60
```

(Columns: mnemonic, Hex. code)

Chapter 6. An Examination of the 8K monitor Program

6.1 The need for a monitor program.

The first point for discussion is whether or not a microcomputer system, such as the ZX-81 system, actually needs to have a monitor program at all. However, it is fairly obvious that when the power is turned on, the Z80 microprocessor will start executing instructions sequentially. In the case of the Z80 the first instruction is collected from location Hex.0000. But it is very important to realise that the microprocessor itself cannot in anyway know whether it is following a sensible machine code program, or just obeying 'rubbish'.

This state can be shown quite nicely by trying to use a ZX-81 from which the 8K ROM has been removed. If this is done, then the user will find that the screen display fails to appear and that the keyboard is inactive.

In order therefore for a microcomputer system to work in an organised manner there must be a monitor program, and for a Z80 microprocessor system the program must be located from Hex.0000 onwards.

The actual size of the monitor program is determined by the number of features that the manufacturer wishes to include in his microcomputer system. For example a hand-held games machine may require a monitor program that occupies less than ½K (512 locations) of memory, whereas a large microcomputer system may have over 100K of memory holding its monitor program.

The standard ZX-81 system is at present supplied with an 8K monitor program. The manufacturer thereby is able to offer a system which can produce and maintain a T.V. display which has 64 different characters, scan a keyboard with 78 different keystrokes and give the user the BASIC language.

In the future it is likely that there will be a range of monitor programs availablesome of which may be smaller and therefore more elementary, and some which are larger and offer enhanced facilities.

6.2 A first look at the structure of the 8K monitor program.

The 8K monitor program can be divided into the following parts.

0000	The RST routines.
007E	The character tables.
0207	The display routines.

02F6	The SAVE and LOAD command routines.
03CB	The initialisation routine.
0419	The BASIC line editing routines.
0C29	The BASIC command tables.
0CBA	The BASIC line scanning routine.
0DAB	The BASIC command routines.
0F52	The BASIC expression evaluator.
1586	The floating-point handling routines.
1914	The function table.
199C	The floating-point calculator.
1E00	The character generator.
1FFF	end.

The 8K monitor program can therefore be seen to be made up of 14 significant parts. In general the routines involved with the display are situated near the beginning of the program, the BASIC interpreter in the middle and the floating-point routines towards the end. The 8K ROM also holds the 'formats' of the 64 display characters in the character generator that occupies the top ½K.

Before discussing these different parts of the monitor program it is worth while considering a functional model of the ZX-81.

6.3 A function model of the ZX-81.

When the power is turned on the first routine to be performed is not, surprisingly, the 'initialisation routine'. However, after that the ZX-81 enters a loop which is comprised of the 'keyboard scanning routine' and the main 'display routine'. The computer stays in this program loop until a key is pressed on the keyboard.

When this occurs an exit is made from the loop and some action is taken. This action leads to the execution of those routines associated with the keystroke. The routines may involve 'editing the display', 'entering a character to the E-line' or 'running a BASIC program'.

Once the appropriate routines have been executed the computer returns to its 'keyboard and display loop', where the whole procedure can be repeated.

The above description applies very well to the 'fast' mode of operation of the ZX-81. However the operation of the 'slow' mode is more complicated. The most simple view possible is to consider that when the Z80 is executing any routine outside the 'keyboard and display loop' there are repeated 'interrupts' on the NMI line. Each interrupt leads to the execution of a 'display routine'.

The following diagram shows these operations.

Diagram 10 A functional model of the ZX-81.

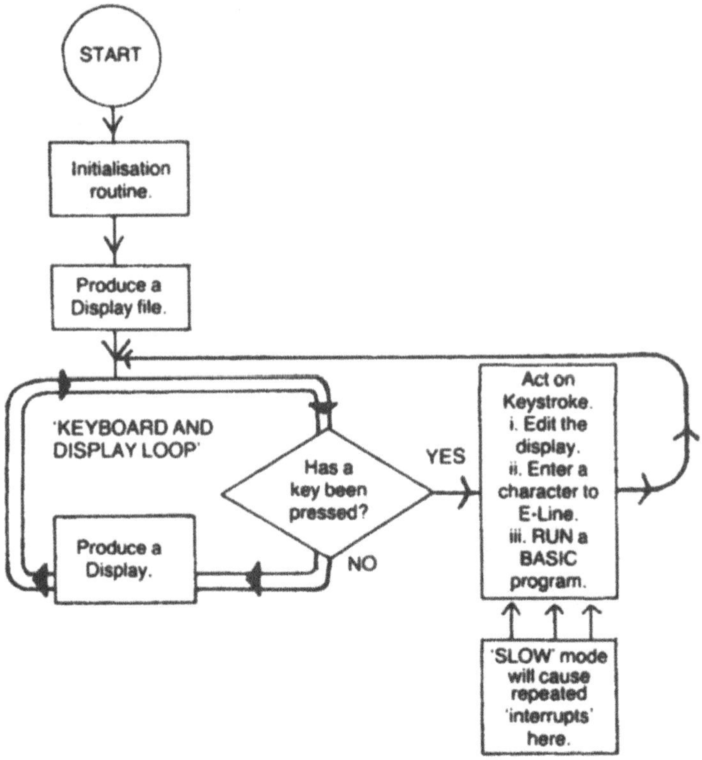

4 The different parts of the 8K monitor program.

ch of the 14 parts of the monitor program will now be discussed in turn. e parts of program that are useful to a programmer writing his own ›grams will be given more explanation than the parts that cannot easily used.

e RST routines:

e RST locations are used for the following purposes:

00	Start.
08	Report handling.
10	Print a character.
18	Collect a character from a BASIC line.
20	Collect next character from a BASIC line.
28	Jump to floating-point calculator.
30	Make space in memory.
38	IM1 interrupt routine for each display line.
&	
66	NM interrupt for 'slow' display routine.

;T locations 0008 and 0010 are worth discussing further.

'RST 0008'

When a report is required a 'RST 0008' instruction is used. The value of the report is required as data in the following byte. The operation of the report handling facility can be shown using the following BASIC program.

	mnemonic	Hex. code
10 SLOW		
20 POKE 17000,207	RST 0008	CF
30 POKE 17001,34		22
40 LET K=USR 17000		
50 PRINT "LINE 50"		

When the above program is RUN it will be seen that "LINE 50" is never printed as report 'Z' has occurred in line 40.

The actual lines from the monitor program that are involved are:

address	Hex. code	mnemonic	comment
0008	2A 16 40	LD HL,(4016)	Save CH-ADD
00B	22 18 40	LD (4018),HL	in X-PTR
00E	18 46	JP 0056	Jump.
. . .			
056	E1	POP HL	Get data byte address off stack
057	6E	LD L,(HL)	Transfer to L.
058	FD 75 00	LD (4000),L	To report code.
05B	ED 7B 02 40	LD SP,(4002)	Prepare to clear stack.
05F	. . . . . .		Clear stack & return.

A programmer may find it useful therefore to use 'RST 0008' in his programs to define his own type of errors.

'RST 0010'

When a character is to be printed the most common method used is to load the A register with the appropriate character code and use a 'RST 0010' instruction. The example from the 8K monitor program on page 64 shows this being done.

The actual lines of the RST 0010 routine are:

address	Hex. code	mnemonic	comment
0010	A7	AND A	Is it a space?
011	C2 F1 07	JP NZ,07F1	No. Print a character.
014	C3 F5 07	JP 07F5	Yes. Print a space.

This routine will always print a character in the next place in the Display File, and it is up to the programmer to ensure that the values of S-POSN, the column number and the line number for the PRINT command, are correct.

The Character tables:
There are four tables in this part of the 8K monitor program. They are:

007E — 00CB The keyboard character table.
00CC — 00F2 The function character table.
00F3 — 0110 The graphic character table.
0111 — 01FB The command character table.

The following simple BASIC program shows these tables:

```
10 SLOW
20 FOR A=126 TO 507
30 IF A=204 OR A=243 OR A=273
THEN PRINT ,,,
40 PRINT CHR$ PEEK A;
50 NEXT A
```

The program gives the 'CHR$' representation of each entry in the table.

The editing commands however come to be displayed as question marks, as do the keystrokes that change the operating modes.

The Display routines:
The routines from 0207-02BA are concerned with the production of a display.

The routine from 0207-0228 is used to determine whether the 'slow' or 'fast' mode is being used and the routine from 0229-0291 is the 'main display routine'.

The following BASIC program can be used to demonstrate this second routine.

```
10 PRINT "LINE 10"
20 FAST
30 LET K=USR 553    Hex. 0229
RUN
```

The effect of using 'USR 553' is to call the 'main display routine' and hence the display file is displayed in an 'unfinished' state. Note that the report message is not printed. However if a key is pressed then an exit is made from the routine and the display is completed and displayed.

The main display routine is in itself made up of three major parts:

i. Decrease and test PAUSE counter.
ii. Scan keyboard.
iii. Produce the display.

The reader will therefore see that it is this routine that is used by the PAUSE command to hold the display for a specified period, and that a

PAUSE period is ended by exiting from this routine by making a keystroke.

The routine at 0292-02B4 is the 'slow' mode display controlling routine, and the routine at 02B5-02BA is the routine that actually starts the display of a single line of characters.

The 'keyboard scanning routine' is a very important routine and is to be found at 02BB-02E6. This routine can be used by the programmer if he wishes as is demonstrated in the following BASIC program.

```
                                        mnemonic      Hex. code
 10 REM 123456
 20 PRINT "AFTER NEWLINE
   HOLD DOWN A KEY"
 30 INPUT A$
 40 CLS
 50 LET A=16514
 60 POKE A,205                          CALL dddd     CD
 70 POKE A+1,187                                      BB
 80 POKE A+2,2                                        02
 90 POKE A+3,68                         LD B,H        44
100 POKE A+4,77                         LD C,L        4D
110 POKE A+5,201                        RET           C9
120 FOR A=1 TO 100
130 NEXT A
140 PRINT "KEY VALUE
    = ";USR 16 514
RUN
```

Note that the key value returns from the routine in the HL register. (see appendix iv. on page 162 for the complete 'Table of Key Values'.

The key values represent the 79 unique codes that are derived from the scanning of the keyboard. In the keyboard decoding routine at 07BD-07DB these values are first changed to an ordered sequence, 1-78, and the character code obtained by referring to the character tables. 'No key pressed' being detected beforehand. (see also page 38)

In a machine code program it is quite practical to call this routine and test the resulting contents of the HL pair against the key values, thereby detecting whether or not specific keys are being pressed. (see page 140) (see page 153 for a full listing of the 'Keyboard scanning routine')

The SAVE and LOAD command routines:

The main part of the save routine is at 02F6-033F, and the LOAD routine from 0340-03A1.

In the ZX-81 system programs saved on cassette tape must have program names. These names are to be found on tape before the actual variables and program. As usual the end of a word, or name, is marked by having the last letter 'inverted'.

In the case of the SAVE command routine there is the initial pause, 02FC-030A, followed by the sending of the program name, 030B-0312, and finally the sending of the system variables, the program and the program variables. Note that the display file is also copied to the tape.

The following lines show some of these features.

In FAST mode enter:

LET A=USR 787

After pressing NEWLINE the program will be SAVEd directly.

LET A=USR 763

and after pressing NEWLINE there will be the familiar 5 second pause before the SAVEing of the program.

In the case of the LOAD command the program name is read in first, followed by the program.

The use of the following line shows that this routine can be entered directly also.

In FAST mode enter:

LET A=USR 839

and the familiar 'waiting for data' pattern will appear.

A full listing of the 'SAVE command routine' and the 'LOAD command routine' are given in appendix i on page 150.

The Initialisation routine:

The different parts of the Initialisation routine are:

		see pages
i.	The RAM check.	53
ii.	The setting of the stack marker.	43, 51
iii.	The loading of the I register.	32
iv.	The selection of interrupt mode 1.	92
v.	The loading of the IY register pair.	31
vi.	The selection of 'slow' mode.	44
vii.	The building of the basic display file.	71

as all these different parts of the routine have been used as examples in Chapter 4 they will not be discussed further.

The BASIC line editing routines:

There are a tremendous number of different routines in this part of the monitor program.

The following list gives the locations of the more important of them. However only the routines that are useful to the machine code programmer will be discussed further.

Ø454 Cursor down routine
Ø482 Build-up an E-line routine
Ø4C1 Command point routine
Ø52B Editing key sort routine.
Ø5C4 Edit routine

Ø63E	Run BASIC program routine
Ø72C	LIST command routine
Ø745	Print a whole BASIC line routine.
Ø7BD	Keyboard decode routine
Ø7F1	Print a character routine
Ø8F5	Test PRINT AT parameters routine
Ø918	Expand Display File routine
Ø94B	Print keywords routine
Ø9AD	Change all pointers routine
Ø9F2	Next variable or BASIC line routine
ØA2A	CLS command routine
ØA98	Print a decimal number routine
ØACF	PRINT command routine
ØBAF	PLOT and UNPLOT commands routines
OCOE	SCROLL command routine

The routines from this part of the monitor that can easily be used in a machine program will now be discussed. Simple BASIC programs will demonstrate the routines being used.

The CLS command routine.

This is a very simple routine to use. In BASIC in order to clear the screen a line such as: 1Ø CLS is used. When the BASIC interpreter executes this line the only action taken is to call the subroutine at 0A2A.

Hence whenever a machine code program contains the instruction line: CALL 0A2A the screen will be cleared.

This is shown in the following BASIC program:

```
                                   mnemonic      Hex. code
 1Ø  REM 1234
 2Ø  PRINT "PRESS NEWLINE
     TO CLEAR SCREEN"
 3Ø  SLOW
 4Ø  LET A=16514
 5Ø  POKE A,2Ø5                    CALL addr.    CD
 6Ø  POKE A+1,42                                 2A
 7Ø  POKE A+2,1Ø                                 ØA
 8Ø  POKE A+3,2Ø1                  RET           C9
 9Ø  INPUT A$
1ØØ  LET L = USR A
RUN
```

The print a decimal number routine:

Although the 8K monitor program provides floating-point arithmetic there is still the need to have a routine for printing decimal numbers. This routine is used to print the contents of the HL register pair as a decimal number using absolute binary arithmetic. The range of numbers printed is 0 to 9999.

The routine is used in the monitor program for the printing of the line numbers in a listing of a BASIC program.

The following BASIC program demonstrates the routine.

```
                                        mnemonic       Hex. code
  10 REM 1234567
  20 SLOW
  30 PRINT AT 18,0;"ENTER A
     VALUE FOR THE HIGH
     BYTE (0-39)"
  40 INPUT H
  50 PRINT AT 18,22;"LOW
     BYTE","(0-255)"
  60 INPUT L
  70 CLS
  80 LET A=16514
  90 POKE A,33                          LD HL,+dddd    21
 100 POKE A+1,L                                        L
 110 POKE A+2,H                                        H
 120 POKE A+3,205                       CALL addr.     CD
 130 POKE A+4,171                                      AB
 140 POKE A+5,10                                       0A
 150 POKE A+6,201                       RET            C9
 160 PRINT "THAT GAVE";
 170 LET L=USR A
 180 RUN
RUN
```

Note how the USR statement cannot this time be included in a PRINT statement.

The PRINT AT routines:
The BASIC PRINT AT command allows the user of a ZX-81 to move the PRINT position to a specified location.

e.g. 10 PRINT AT 10, 10

will lead to the PRINT position being the eleventh space on the eleventh line. (do not forget that 0,0 is the top lefthand space)

When the display file is in a 'collapsed' state the use of the PRINT AT command results in the display line being expanded if it should be necessary. If the specified space already exists in the Display File then no expansion is required.

In the 8K monitor program whenever a PRINT AT command is executed, the parameters supplied by the programmer are tested in the 'test PRINT AT parameters routine' (see also page 53) and then the 'expand Display File routine' is called. This second routine tests to see whether or not the specified PRINT position already exists in the Display File and if it does not, the Display File is expanded so that it does become present.

Both of these routines can be used in machine code programs. The following BASIC program gives a very simple demonstration of the PRINT position being set by a machine code routine.

```
                                         mnemonic      Hex. code
 10 SLOW
 20 PRINT AT 18,0;"ENTER
    LINE NUMBER (0-16)"
 30 INPUT L
 40 PRINT AT 18,0;"ENTER
    COLUMN"
 50 INPUT C
 60 PRINT AT 18,0;"ENTER
    YOUR CHARACTER NOW"
 70 INPUT B$
 80 LET A=16444
 90 POKE A,1                             LD BC,dddd    01
100 POKE A+1,C                                         C
110 POKE A+2,L                                         L
120 POKE A+3,205                         CALL addr.    CD
130 POKE A+4,245                                       F5
140 POKE A+5,8                                         08
150 POKE A+6,201                         RET           C9
160 LET K=USR A
170 PRINT B$
180 RUN
RUN
```

In the above program the user is asked to provide the line and column parameters, and a character, or string of characters. The machine code routine then sets the required PRINT position for line 170.

The PRINT a string routine.

This routine forms part of the 'PRINT command routine', and is used in the 8K monitor program whenever a string of characters has to be printed.

The parameters for handling a string are:

i. The starting address of the string, and

ii. The number of characters of the string.

These parameters are passed to the 'PRINT a character routine' for each of the characters in the string, and this results in each character being printed at the current PRINT position. Remember that the 'PRINT a character routine' increments the PRINT position after the character has been printed.

The following BASIC program shows a string being printed.

```
                                      mnemonic Hex. code
10  REM THIS STRING HAS
     29 CHARACTERS
20  SLOW
30  PRINT AT 8,0;
40  LET A=16444
50  POKE A,17                         LD          11
                                      DE,+dddd
60  POKE A+1,130                                  82
70  POKE A+2,64                                   40
80  POKE A+3,1                        LD          01
                                      BC,+dddd
90  POKE A+4,29                                   1D
100  POKE A+5,0                                   00
110  POKE A+6,205                     CALL addr   CD
120  POKE A+7,107                                 6B
130  POKE A+8,11                                  0B
140  POKE A+9,201                     RET         C9
150  LET L=USR A
     RUN
```

In the above program the DE register pair is loaded with the starting address of the string – decimal 16514, Hex. 4082, and the BC registers pair is loaded with the number of characters in the string – decimal 29, Hex. 001D.

The PLOT and UNPLOT commands routine.

This routine is used in a very similar way to the 'PRINT AT routine'.

Initially therefore the parameters passed to the routine are tested to ensure that they are within the correct range.

A test is then made to see if the operation should be a PLOT or an UNPLOT. This test is performed by comparing the value of T-ADDR, the system variable that holds the address of the next item in the syntax table, against the address in that table for the UNPLOT command.

The following lines show the actual test.

address	Hex. code	mnemonic	comment
0BDA	11 9E 0C	LD DE,+0C9E	UNPLOT address in table.
BDD	3A 30 40	LD A,(T-ADDR)	Pick-up T-ADDR.
BE0	93	SUB E	Test low bytes.
BE1	FA E9 0B	JP M,0BE9	Jump for PLOT

BE4 . . . . . continue with UNPLOT.

The following BASIC program shows how the operations of PLOT and UNPLOT can be included in a machine code routine.

```
                                        mnemonic      Hex. code
 10 SLOW
 20 PRINT AT 18,0;"ENTER X
    PIXEL NUMBER (0-63)"
 30 INPUT X
 40 PRINT AT 18,6;"Y PIXEL
    NUMBER (10-43)"
 50 INPUT Y
 60 PRINT AT 18,0;"ENTER PLOT
    OR UNPLOT (P/U)"
 70 INPUT B$
 80 LET A=16444
 90 POKE A,1                            LD BC,dddd    01
100 POKE A+1,X                                        X
110 POKE A+2,Y                                        Y
120 POKE A+3,62                         LD A,dd       3E
130 POKE A+4,102+CODE
B$                                                    9B or A0
140 POKE A+5,50                         LD (addr.),   A, 32
150 POKE A+6,48 (T-ADDR)                              30
160 POKE A+7,64 (          )                          40
170 POKE A+8,205                        CALL addr.    CD
180 POKE A+9,178                                      B2
190 POKE A+10,11                                      OB
200 POKE A+11,201                       RET           C9
210 LET K=USR A
220 RUN
RUN
```

In the above program the user is asked to provide valid parameters for X and Y, and then to specify PLOT or UNPLOT. The value stored as the low byte of T-ADDR is then compared to the constant Hex.9E to distinguish whether the operation is to be a PLOT or an UNPLOT.

The BASIC command tables:

The first table in this part of the monitor program is the table of 'offsets' that is used to index into the 'syntax' table.

The offset table' is at 0C29 — 0C47, and the 'offsets' for the different BASIC commands can be shown by using the following BASIC program.

```
 10 SLOW
 20 FOR A=1 TO 2
 30 PRINT "COMMAND OFFSET";
 40 NEXT A
 50 PRINT
 60 PRINT
 70 LET B=8
 80 FOR A=225 TO 255
 90 PRINT CHR$ A;TAB (16-B);
    PEEK (2888+A),
100 LET B=-B
110 NEXT A
RUN
```

The other table in this part of the monitor program is the 'syntax' table.

This table gives the command class for each part of the command, the character code for the required syntactic separator, if one is required, and the address of the command routine.

e.g. The entry for PLOT is:

0C98 06	The command class.
0C99 1A	The separator, a 'comma'.
0C9A 06	The command class.
0C9B 00	The command class.
0C9C AF)	The PLOT routine is at
0C9D 08)	OBAF which determines that a PLOT command line must have the syntax;
PLOT	(expression) (,) (expression) as in PLOT 0,0

The following table gives the command routine addresses as can be obtained from the table.

Address	Command	Command routine address
0C48	LET	(not in table but it is 131D)
0C4B	GOTO	0E81
0C4F	IF	0DAB
0C54	GOSUB	0EB5
0C58	STOP	0CDC
0C5B	RETURN	0ED8
0C5E	FOR	0DB9
0C66	NEXT	0E2E
0C6A	PRINT	0ACF
0C6D	INPUT	0EE9
0C71	DIM	1405
0C74	REM	0D6A
0C77	NEW	03C3
0C7A	RUN	0EAF
0C7D	LIST	0730
0C80	POKE	0E92
0C86	RAND	0E6C
0C89	LOAD	0340
0C8C	SAVE	02F6
0C8F	CONT	0E7C
0C92	CLEAR	1496
0C95	CLS	0A2A
0C98	PLOT	0BAF
0C9E	UNPLOT	0BAF
0CA2	SCROLL	0C0E
0CA7	PAUSE	0F2F
0CAB	SLOW	0F28
0CAE	FAST	0F20
0CB1	COPY	0869
0CB4	LPRINT	0ACB
0CB7	LLIST	072C

The BASIC line scanning routine.

This part of the 8K monitor program is the real BASIC interpreter. It is in this routine that each BASIC line is scanned and actions determined that are needed to execute that line.

As a BASIC line is scanned each command is identified. The command class for that command is obtained from the syntax table and a call made to the appropriate command class routine.

The different classes are used to show the differing syntactic requirements of the commands.

e.g. The STOP command is in class 0, as it is a command that cannot be followed by any parameters. Whereas PLOT is put into class 6 on two occasions, as the PLOT command is required to be followed by two expressions separated by a 'comma'.

The BASIC command routines:
Each of the BASIC commands has a command routine in the 8K monitor program. Most of the routines are to be found in this block of the program. The addresses for all of the routines are to be found in the table on page 123.

The majority of the routines are too complicated to be discussed, but it is worth looking at the FAST command routine and the SLOW command routine as these routines can be used by the machine code programmer.

The FAST command routine:
The full routine is:

address	**Hex. code**	**mnemonic**	**comment**
0F20	CD E7 02	CALL 02E7	See below.
F23	FD CB 3B B6	RES 6,(403B)	Reset the Flag.
F27	C9	RET	Finished.
02E7	FD CB 3B 7E	BIT 7,(403B)	FAST already?
2EB	C8	RET Z	Yes
2EC	76	HALT	No. So wait interrupt.
2ED	D3 FD	OUT (+FD),A	To the logic chip.
2EF	FD CB 3B BE	RES 7,(403B)	Reset the other flag.
2F3	C9	RET	Finished.

In a machine code program the programmer wishes to change to FAST mode, or to ensure that he is in FAST mode, then it can be done simply by using: CALL 0F20.

The SLOW command routine:
The routine is only:

address	**Hex. code**	**mnemonic**	**comment**
0F28	FD CB 3B F6	SET 6,(403B)	Set the flag.
F2C	C3 07 02	JP 0207	Jump to display routine.

This routine can be called using: CALL 0F28.

The BASIC expression evaluator and the floating-point routines.

These parts of the 8K monitor program are very complicated and cannot be used by a machine code programmer.

The character generator:

The following BASIC program can be used to show the 64 different characters in the 'character generator', each enlarged 64 times.

```
10  FAST
20  FOR A=7680 TO 8184 STEP 8
30  PRINT "ADDRESS CONTENTS CHARACTER"
40  PRINT AT 3,17;"************"
50  FOR B=A TO A+7
60  LET C=PEEK B
70  PRINT B;TAB 8;C;TAB 17;"*";
80  LET D=128
90  FOR E=0 TO 7
100 IF C>D-1 THEN GOTO 130
110 PRINT " ";                              (space)
120 GOTO 150
130 PRINT "■";                              (graphic space)
140 LET C=C-D
150 LET D=D/2
160 NEXT E
170 PRINT " *"
180 NEXT B
190 PRINT AT 12,17;"************"
200 PAUSE 150
210 CLS
220 NEXT A
RUN
```

Chapter 7. Using Machine Code Routines in BASIC Programs.

7.1 Introduction

The main purpose of using machine code routines within a BASIC program is to produce a final program that runs considerably faster than the original program.

It is rarely practical to completely replace the whole of a BASIC program with a single machine code program in the ZX-81 system using the standard 8K ROM. The main problem being that there is not a particular subroutine in the 8K monitor program for handling the 'INPUT command'. Therefore most programmers write machine code routines for the slow parts of their programs and return to BASIC whenever an 'INPUT' is required.

The resultant program could therefore be of the form:

```
10 REM xxxxxxxxxxxxxxxxxxxxxxx   ) the machine code.
xxxxxxxxxxxxxxxxxxxxxxxxxxxxx    ) routines.
20 LET K=USR 16514               Execute the first routine.
30 INPUT A                       Return for INPUT.
40 POKE 16530,A                  Store the INPUT.
50 LET K=USR 16531               Execute the next routine.
RUN
```

The rest of this chapter takes the reader through the actual stages in producing 'machine code routines'.

7.2 The stages involved.

The first stage is the 'writing of the BASIC program'.

It is usually a very good idea to produce a BASIC program that actually performs the required task. Of course the program may run very slowly but at least the programmer has the opportunity to make changes relatively easily.

The second stage is the 'deciding which part of the program is to be converted to machine code'.

This stage involves determining which set of BASIC lines are to be replaced by a: LET K=USR xxxxx and also which parameters are required by the machine code routine, on entry; and which parameters are to be returned by the routine.

The third stage is the 'producing of a listing of the machine code routine in assembler format'.

This stage is usually the most difficult as it involves the conversion of the original algorithm to one that uses the rather limited number of registers within the Z80. The programmer also has to consider which instructions in the Z80 instruction set actually exist, and the 'constructions' involved for those operations that are not suported by the Z80.

e.g. There is no instruction — LD B,(addr.) and the usual construction to perform this would be — LD A,(addr.) LD B,A but this disturbs the A register. An alternative construction would be — LD BC,(addr. − 1) but this disturbs the C register and a third construction would be — LD HL,+dddd (address) LD B,(HL) that disturbs the HL register pair.

The fourth and final stage is the 'assembling of the machine code routine'.

This stage involves the actual production of the appropriate Hex. code or decimal code for the routine. The location of the program in the RAM will usually affect this operation.

A machine code routine that can be placed at any point in the RAM because it does not contain any absolute addresses is said to be 'relocatable'. However most routines do contain absolute addresses and therefore must be 'located' in a particular part of the RAM.

The following examples show these stages in detail.

7.3 The first example — The Sum of a Series.

This example was briefly discussed on pages 99 & 100 but the stages involved in producing the machine code routine were not dealt with in detail.

The following BASIC program asks the user to enter an integer in the range 1-255 and then produces: THE SUM OF THE SERIES 0+1+2+. . . N = xxx.

Stage 1.

The program for this general case is:

```
 20 SLOW
 30 PRINT AT 18,Ø;"ENTER A VALUE
    FOR N (1-255)"
 40 INPUT N
 50 CLS
 60 PRINT "THE SUM OF THE SERIES"
 70 PRINT AT 2,2;"Ø+1+2+. . .";N;" = ";
 80 LET SUM=Ø
 90 FOR K=N TO Ø STEP -1
100 LET SUM=SUM+K
110 NEXT K
120 PRINT SUM
130 RUN
```

This program satisfies the requirement of 'stage 1' in that a working BASIC program has been produced.

Notice how the program has been written.

It is obviously going to be desirable to replace lines 80-110 with machine code, therefore these lines have been written like a 'subroutine' at the end of the program. Note also how the 'FOR . . .NEXT' loop goes from '255. . .to. . .0' which will make it very easy to convert it to a DJNZ instruction.

The more 'machine code features' that are included in the BASIC program, then generally the easier it will be to convert the program to machine code.

Stage 2.

In this simple example program it is lines 80-110 that are going to be replaced by a machine code routine.

The variable N, range 0-255, is the only parameter required by the routine, and the variable SUM is the only parameter returned by the routine. In this case the lines:

```
80 POKE 16514,N
90 PRINT USR 16515
```

can be used to replace lines 80-120.

It is useful at this stage to produce a flow diagram of the actual lines that are to form the machine code routine.

A possible flow diagram is:

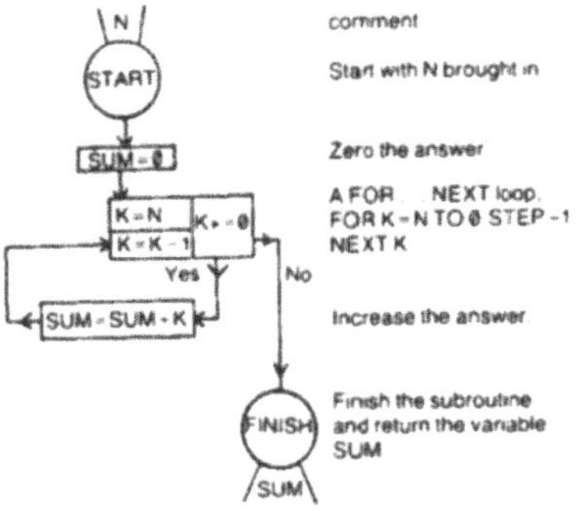

Stage 3.

The first task in this stage is to make a start at allocating 'registers' or 'memory locations' to the various variables.

The flow chart developed for stage 2 shows that the FOR. . .NEXT loop box is central to the flow diagram. Hence it would be sensible to assign a register to the loop counter K.

In this example the B register used in conjunction with a DJNZ instruction can very easily be used as a loop counter.

The FOR. . .NEXT loop box can now be redrawn as:

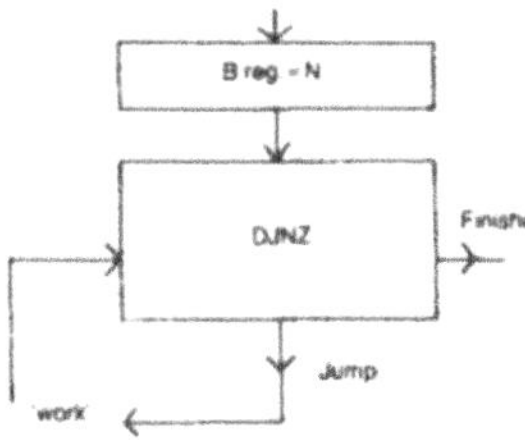

Note that the B register has to be initialised to the value of N.

There are many ways of dealing with the variable SUM but probably the most simple method is to allocate the HL register to this variable.

Initially HL will have to be zeroed, and at the end of the routine the contents of the HL register will have to be copied into the BC register for returning as the USR variable.

Note that the BC register pair cannot be used to hold SUM throughout the routine as the B register has already been used.

The operation of 'SUM=SUM+K' is not possible directly as there is no instruction 'ADD HL,B' therefore a 'construction' has to be considered. As there is a suitable instruction 'ADD HL,DE' it would be quite sensible to use it and to copy the contents of B into E before the addition operation. However the D register must be zeroed at the start of the routine.

The last point to be made is that a 'construction' must be used for the initialising of the B register to hold the variable N.

The use of: LD A,addr. and LD B,A is quite appropriate.

The full flow chart now becomes as follows. The mnemonics are also given:

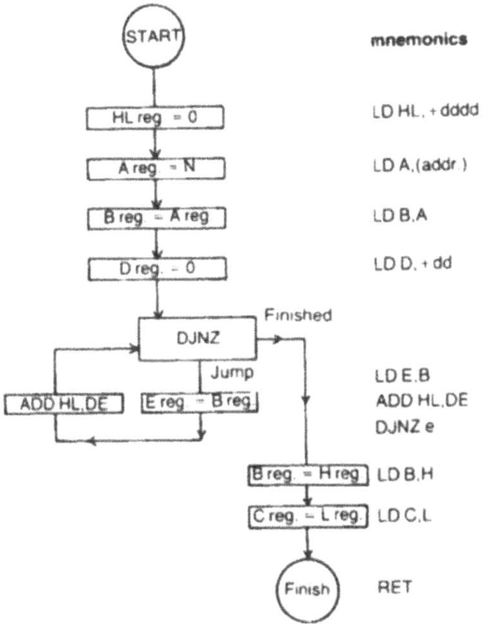

Stage 3 requires that a listing of the final machine code routine be given in 'assembler format', and therefore the routine becomes:

Label	**mnemonic**
START	LD A,(N)
	LD B,A
	LD HL,+0000
	LD D,+00
LOOP	LD E,B
	ADD HL, DE
	DJNZ LOOP
	LD B,H
	LD C,L
END	RET

Stage 4

The variable N is going to be POKEd into location 16514 and the routine located from 16515 onwards. The 'assembled' routine will be:

address. dec.	Hex.	Hex. code	Dec. code	label	mnemonic
16514	4082	—	—	N	LD A,(N)
16515	4083	3A	58	START	LD A,(N)
16516	4084	82	13Ø		
16517	4085	40	64		
16519	4087	21	33		LD HL,+0000
16520	4088	00	Ø		
16521	4089	00	Ø		
16522	408A	16	22		LD D,+00
16523	408B	00	Ø		
16524	408C	58	88	LOOP	LD E,B
16525	408D	19	25		ADD HL,DE
16526	408E	10	16		DJNZ LOOP
16527	408F	FC	252		
16528	4090	44	68		LD B,H
16529	4091	4D	77		LD C,L
16530	4092	C9	201	END	RET

The table shows that the final routine uses 17 locations.

Now comes the moment of truth. Enter the following BASIC lines:

```
10  REM 12345678901234567  The 17 locations.
20  SLOW
30  PRINT AT 18,0;"ENTER A VALUE
    FOR N (1-255)"
40  INPUT N
50  CLS
60  PRINT "THE SUM OF THE SERIES"
70  PRINT AT 2,2;"0+1+2+. . .";N;" = ";
80  POKE 16514,N
90  PRINT USR 16515
100  RUN
```

Do not RUN the program until you have entered the machine code routine into line 10.

A useful method of 'loading machine code' is as follows:

Enter:

```
200 FOR A=16515 TO 16530
210 INPUT B
220 POKE A,B
230 NEXT A
RUN 200
```

This routine is RUN and the decimal code entered. In this case it is: 58,130,64,71,33,00,22,0,88,25,16,252,68,77,201.

Note that location 16514 is left to be filled by the BASIC program.

Once the machine code has been entered, delete lines 200-230 and enter RUN.

Although this program is only very simple, there is an impressive change in the response times, when compared to the BASIC-only version.

7.4 The second example – A Moving Ball Program.

This example is a good deal more complicated than the first example but it has been included as it shows how a complicated program can be tackled.

Stage 1.

The BASIC program is:

```
20 SLOW
30 CLS
40 GOSUB 100
```

```
 50 FOR A=1 TO 16
 60 PRINT AT A,0;"■"; AT A,18;"■"       graphic space
 70 NEXT A
 80 GOSUB 100
 90 GOTO 140
100 FOR A=0 TO 18
110 PRINT "■";                           graphic space
120 NEXT A
130 RETURN
140 LET LR=1
150 LET UD=1
160 LET X=2
170 LET Y=INT (RND*32+10)
180 PLOT X,Y
190 IF INKEY $="R" THEN RUN
200 IF X=2 OR X=35 THEN LET LR=-LR
210 IF Y=10 OR Y=41 THEN LET UD=-UD
220 UNPLOT X,Y
230 LET X=X-LR
240 LET Y=Y-UD
250 GOTO 180
RUN
```

The lines 20-130 draw a rectangular playing area. The use of a temporary line '135 STOP' can be used to show this being done.

Lines 140 and 150 initialise the 'direction' of the ball. LR is the variable for 'Left and Right', +1 is for moving right and −1 is for moving left. UD is the variable for 'Up and Down', +1 is for moving up and −1 is for moving down.

Lines 160 and 170 give initial values to the X and Y variables used in the PLOT command. The value for X is fixed but the value for Y is chosen at random.

Line 190 allows the user to restart the program.

Lines 200 and 210 test the position of the ball and if it is against an edge the direction of the ball is reversed.

Line 220 erases the position that was filled in line 180.

Lines 230 and 240 give new values for X and Y so that the ball moves.

Stage 2.

The functional flow diagram for this program could be:

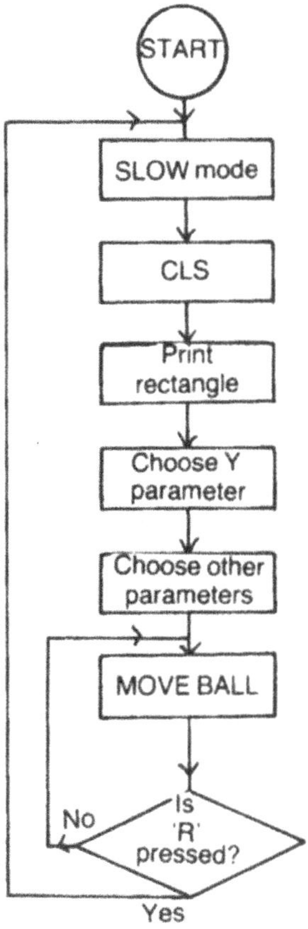

It is possible to write machine code, of quite a simple nature, for all of the functional blocks of this program with the exception of the 'Choose Y parameter' block. This block uses the BASIC RND command for which there is no readily useable subroutine. (The RND command produces a 5 byte floating-point number).

A suitable plan for the final BASIC program would be:

```
10 REM  xxxxxxxxx etc.                The machine code.

20 LET K=USR  xxxxx                   ( SLOW mode
                                      ( CLS
                                      ( Print rectangle
30 POKE xxxxx,  INT(RND*32+10)        ( Choose Y parameter.
40 LET K=USR xxxxx                    ( Choose other  parameters.
                                      (MOVE BALL
                                      (Is 'R' pressed?)

50 RUN
```

However further consideration should be given to 'exiting' from the program. When a BASIC program is being executed the BREAK key is tested after each BASIC line has been dealt with, but in a machine code routine the BREAK key will be inactive unless special provision is included in the routine. In this example, therefore, the keys 'R' and 'BREAK' should be tested.

The flow diagrams for the two machine code routines are as follows:

The first machine code routine: The second machine code routine:

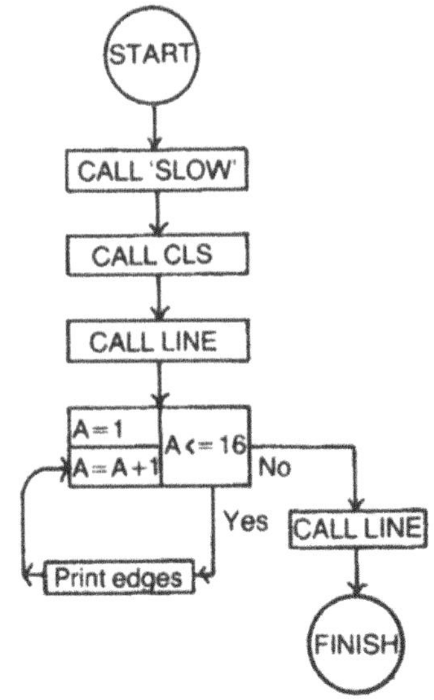

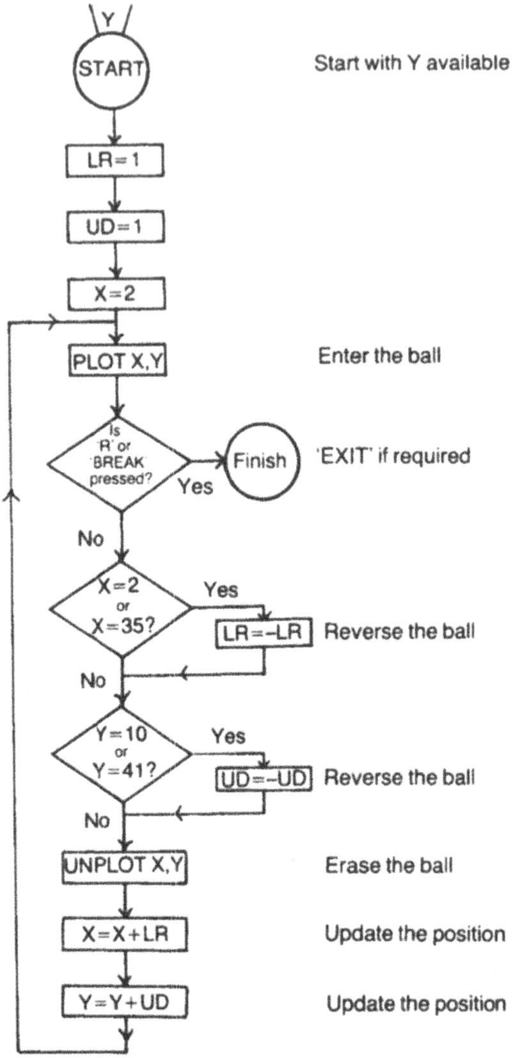

The 'LINE' subroutine.

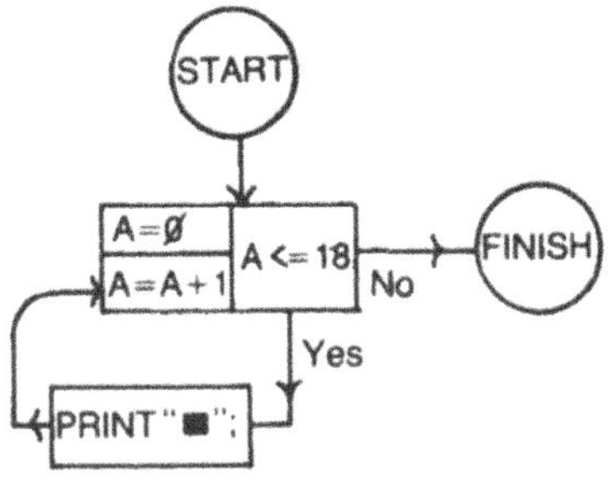

Stage 3:
The first machine code routine;

It is usually easiest to start with the subroutines and then return to the main routine.

The subroutine 'LINE' is a very simple printing operation that prints nineteen identical characters. The use of a DJNZ instruction is most appropriate. The 'assembler format' listing for this subroutine is therefore:

Label	**mnemonic**	**comment**
LINE	LD A,+80	An inverse space.
	LD B,+13	The nineteen characters.
NEXT	RST 0010	Print a character.
	DJNZ NEXT	Go back.
	RET	Finished.

In the main routine the first two CALLs are to routines within the 8K monitor program and the 'assembler format' listing for the routine, except for the 'print edges' block will be:

Label	**mnemonic**
START	CALL SLOW
	CALL CLS
	CALL LINE
EDGES	....
	CALL LINE
	RET

The conversion of the 'print edges' block is a little complicated. However it can be performed in simple stages.

The method discussed on page 130 for the first example program, involved drawing a flow diagram where each block corresponded to a machine code instruction. However there is an alternative method that involves the production of a routine of BASIC lines, with each line having a corresponding machine code instruction.

This second method is demonstrated using the three lines of the 'print edges' block.

The original lines are:

```
50 FOR A=1 TO 16
60 PRINT AT A,0;"■"; AT A,18;"■"
70 NEXT A
```

As a first step the loop counter should be reversed, as the final machine code routine will use a DJNZ instruction to operate the looping operation.

Next the 'PRINT AT' command should be split into smaller parts.

The routine will now be:

```
50 FOR B=16 TO 1              Note B is used
STEP -1
54 PRINT AT 17-B,0;
58 LET A=128                  Note A is used.
62 PRINT CHR$ A
66 PRINT AT 17-B,18;
70 LET A=128
74 PRINT CHR$ A
78 NEXT B
```

Note how variable names that are also used as register names are being employed whenever the register is appropriate.

The 'PRINT AT command routine' uses the B register to hold the 'line number' and the C register to hold the 'column number' (see page 118), and this point should now be included. However this will lead to B being used twice unless the first value for be is saved. The stack is a suitable place to save the contents of the B register, hence the variable STACK can be used.

The routine now becomes:

```
                              comment
50 FOR B=16 TO 1 STEP -1
52 LET STACK=B                Save B temporarily.
54 LET B=17-B                 Re-define B, - Line.
56 LET C=0                    Define C, - Column.
58 PRINT AT B,C;
60 LET A=128                  Print the 'inverse space'.
62 PRINT CHR$ A
64 LET B=STACK                Collect B.
66 LET B=17-B                 Re-define B, - Line.
68 LET C=18                   Define C, - Column.
70 PRINT AT B,C;
72 LET A=128                  Print the 'inverse space'.
74 PRINT CHR$ A
76 LET B=STACK                Restore B.
78 NEXT B                     Next line.
```

The last stages involve the splitting of the 'B=17−B' and the balancing of the lines involving the stack. This point is very important in machine code as the moving of data onto or off the stack will change the value of the stack pointer. In BASIC however there is no such problem.

The result is that the line: 64 LET B=STACK must be followed by a line: 65 LET STACK=B.

The routine now becomes:

```
                                              mnemonic that corresponds.
EDGES   50 FOR B=16 TO 1 STEP -1     LD B,+dd
        51 LET STACK=B               PUSH BC
        52 LET A=17                  LD A,+dd
        53 LET A=A-B                 SUB B
        54 LET B=A                   LD B,A
        56 LET C=0                   LD C,+dd
        58 PRINT AT B,C;             CALL PRINT AT
        60 LET A=128                 LD A,+dd
        62 PRINT CHR$ A              RST 0010
        64 LET B=STACK               POP BC
        65 LET STACK=B               PUSH BC
        66 LET A=17                  LD A,+dd
        67 LET A=A-B                 SUB B
        68 LET B=A                   LD B,A
        69 LET C=18                  LD C,+dd
        70 PRINT AT B,C;             CALL PRINT AT
        72 LET A=128                 LD A,+dd
        74 PRINT CHR$ A              RST 0010
        76 LET B=STACK               POP BC
        78 NEXT B                    DJNZ LINE 51
```

The routine now has its 'one-to-one' correspondence but it can be improved by re-writing it using a subroutine. It becomes:

```
EDGES   50 FOR B=16 TO 1 STEP -1     LD B,+dd
        51 LET STACK=B               PUSH BC
        52 LET C=0                   LD C,+dd
        53 GOSUB 70                  CALL SQUARE
        54 LET B=STACK               POP BC
        55 LET STACK=B               PUSH BC
        56 LET C=18                  LD C,+dd
        57 GOSUB 70                  CALL SQUARE
        58 LET B=STACK               POP BC
        59 NEXT B                    DJNZ LINE 51
        60 STOP
```

```
SQUARE 70 LET A=17              LD A,+dd
       71 LET A=A-B             SUB B
       72 LET B=A               LD B,A
       73 PRINT AT B,C;         CALL PRINT AT
       74 LET A=128             LD A,+dd
       75 PRINT CHR$ A          RET 0010
       76 RETURN                RET
```

Before the actual listing is given there is just one further point to be made.

The 'CALL LINE' subroutine that forms the bottom line of the rectangular playing area requires that the PRINT position be moved to the start of a new line, after the last edge square has been drawn.

This can be produced by using the 'RET 0010' instruction to PRINT a NEWLINE character. However it is not totally straightforward as the value 118, Hex.76, cannot be used in a REM statement. The following construction is therefore needed.

Label	**mnemonic**
NEWLINE	LD A,+75
	INC A
	RST 0010

The 'assembler format' listing for the whole of the first machine code routine can now be given.

Label	**mnemonic**	**comment**
LINE	LD A,+80	Inverse space.
	LD B,+13	19 characters.
NEXT	RST 0010	
	DJNZ NEXT	
	RET	
SQUARE	LD A,+11	Decimal 17.
	SUB B	
	LD B,A	
	CALL PRINT AT	In monitor at 08F5.
	LD A,+80	
	RST 0010	
	RET	
START	CALL SLOW	In monitor at 0F28.
	CALL CLS	In monitor at 0A2A.
	CALL LINE	
EDGES	LD B,+10	16 rows, numbers 1 to
ROW	PUSH BC	16.

	LD C,+00	Column 0.
	CALL SQUARE	
	POP BC	
	PUSH BC	
	LD C,+12	Column 18.
	CALL SQUARE	
	POP BC	
	DJNZ ROW	
NEWLINE	LD A,+75	
	INC A	Form Hex.76.
	RST 0010	
	CALL LINE	
	RET	

The above routine may appear initially to be rather awkward but when assembled into machine code and run, even in 'slow' mode the playing area is produced in 0.4 seconds.

The reader may if he wishes turn to page 148, enter this routine and see it working, before returning to deal with the second machine code routine (use 210 FOR B=16514 TO 16567 and LET K=USR 16514).

The flow diagram for this routine, given on page134 , shows that the first task is to initialise the variables LR, UD and X.

There are many ways in which variables such as these could be handled in a machine code routine, but the simplest way is to allocate a memory location to each variable, and label the locations appropriately.

The 'assembler format' listing for this part of the routine could therefore be:

Label	**mnemonic**	**comment**
UD	—	
LR	—	
X	—	
Y	—	Filled from BASIC.
VALUES	LD A,+01	
	LD (UD),A	
	LD (LR),A	
	INC A	The A register now holds Hex. 02.
	LD (X),A	

The next stage is to deal with the 'PLOT X,Y' operation. The example program on page 121 shows how this operation can be performed.

The steps are:

i. Load the BC register pair with X and Y. The X value going to the C register and the Y value going to the B register.

This will simply be: PLOT LD BC,(X). Note that this instruction will be a 'register pair' loading instruction, with X going to C, and Y going to B.

ii. Set the system variable, T-ADDR to 'PLOT'. This will be the same as on page 121.

```
            LD A,+9B a suitable constant.
            LD (T-ADDR),A
```

iii. Call the 'PLOT command routine'.

```
            CALL PLOT command routine.
```

The 'Is R or BREAK pressed?' operation is also fairly simple.

A call to the 'keyboard scanning routine' returns a key value in the HL register pair. This value has then to be tested to see whether it is a 'NO KEY', an 'R' or a 'BREAK'.

The following BASIC lines show just one possible way of performing this operation.

```
                                  comment
10 LET HL=-1                      -1 is 'no key',
                                  61435 would be 'R'.
                                  64895 would be 'BREAK'.
                (these values are given in appendix iv.)
20 LET DE=HL                      Exchange registers.
30 LET HL=61435                   Set HL to hold key value for 'R'.
40 LET HL=HL-DE                   Test for 'R'.
50 IF HL=0 THEN RETURN            'RET Z'
60 LET HL=64895                   Key value for 'BREAK'.
70 LET HL=HL-DE                   Test for 'BREAK'.
80 IF HL=0 THEN RETURN            'RET Z'.
90 PRINT "NO KEY PRESSED"
```

The reader might like to try this little program with different values in line 10. An 'ERROR 7' is obtained when a match is found, otherwise the 'NO KEY PRESSED' message is given.

These BASIC lines can now be converted to give an 'assembler format' listing. Note however that 'AND A' instructions must be used to clear the Carry flag before each subtraction.

Label	mnemonic	comment
KEY TEST	CALL KEYBOARD	In monitor at 02BB.
	EX DE,HL	
	LD HL,+EFFB	Key value for 'R'.
	AND A	
	SBC HL,DE	

```
        RET Z            Return if 'R' pressed.
        LD HL,+FD7F      Key value for 'BREAK'
        AND A
        SBC HL,DE
        RET Z            Return if 'BREAK'
                         pressed.
```

An alternative to the above method would be to signal an 'ERROR' when the 'BREAK' key is pressed. This can be done using an 'RST 0008' instruction.

The testing of the X value can be considered as the following BASIC lines

```
10 LET LR=1                   Define LR.
20 LET X=12                   Define X to a suitable value.
30 LET A=X                    LD A,(X)
40 IF A=2 THEN GOTO 60        Test against 2.
50 IF A<>35 THEN GOTO 80      Test against 35.
60 LET A=LR                   LD A,(LR)
70 LET LR=-LR                 2's complement LR.
80 REM CONTINUE
```

The testing of the Y value will be similar, as only the constants and variables are changed.

Note that as the values of LR and UD change from +1 to −1 it is the 2's complement of LR and UD that have to be found if a match occurs.

The 'assembler format' listing for the two test routines will be:

```
Label        mnemonic
X-TEST       LD A,(X)
             CP +02
             JR Z,LR-REV
             CP +23
             JR NZ,Y-TEST
LR-REV.      LD A,(LR)
             CPL
             INC A
             LD (LR),A
Y-TEST       LD A,(Y)
             CP +0A
             JR Z,UD-REV.
             CP +29
             JR NZ,UNPLOT
UD-REV.      LD A,(UD)
             CPL
             INC A
             LD (UD),A
```

The UNPLOT operation is the same as the PLOT operation but the constant loaded into T-ADDR has to be altered.

Label	mnemonic
UNPLOT	LD BC,(X)
	LD A,+A0
	LD (T-ADDR),A
	Call Plot command routine.

The final operation is the updating of X and Y. Once again there are many ways in which this can be done and the following listing shows a method that uses 'indirect addressing'.

Label	mnemonic	comment
X-UPDATE	LD HL,+LR	address of LR.
	LD BC,+X	address of X.
	LD A,(BC)	collect X.
	SUB (HL)	X=X−LR.
	LD (BC),A	restore X.
Y-UPDATE	DEC HL	move to UD.
	INC BC	move to Y.
	LD A,(BC)	collect Y.
	SUB (HL)	Y=Y−UD
	LD (BC),A	restore Y
	JR PLOT	'GOTO 180'

Note that this being the final part of the routine a 'JR PLOT' is now required.

The whole listing for the second machine code routine can now be given.

Label	mnemonic	comment
UD	—	)
LR	—	) 4 locations for
X	—	) the variables.
Y	—	)
VALUES	LD A,+01	
	LD (UD),A	
	LD (LR),A	
	INC A	
	LD (X),A	
PLOT	LD BC,(X)	
	LD A,+9B	
	LD (T-ADDR),A	Location Hex. 4030.
	CALL PLOT c.r.	Location Hex. 0BB2.
KEY TEST	CALL KEYBOARD	Location Hex. 02BB.

```
              EX DE,HL
              LD HL,+EFFB       'R' key.
              AND A
              SBC HL,DE
              RET Z
              LD HL,+FD7F       'BREAK' key.
              AND A
              SBC HL,DE
              RET Z
X-TEST        LD A,(X)
              CP +02            Lefthand side.
              JR Z,LR-REV.
              CP +23            Righthand side.
              JR NZ,Y-TEST
LR-REV.       LD A,(LR)         Reverse the ball.
              CPL
              INC A
              LD (LR),A
Y-TEST        LD A,(Y)
              CP +0A            Bottom of area.
              JR Z,UD-REV
              CP +29            Top of area.
              JR NZ,UNPLOT
UD-REV        LD A,(UD)         Reverse the ball.
              CPL
              INC A
              LD (UD),A
UNPLOT        LD BC,(X)         Erase the ball
              LD A,+A0
              LD (T-ADDR),A
              CALL PLOT c.r.    Really 'UNPLOT c.r.'
X-UPDATE      LD HL,LR          New X value.
              LD BC,+X
              LD A,(BC)
              SUB (HL)
              LD (BC),A
Y-UPDATE      DEC HL            New Y value.
              INC BC
              LD A,(BC)
              SUB (HL)
              LD (BC),A
              JR PLOT           Round again.
```

Stage 4
The 'assembled' routine will be:

THE FIRST MACHINE CODE ROUTINE

address. dec.	Hex.	Hex. code	Dec. code	Label	mnemonic
16514	4082	3E	62	LINE	LD A, +80
16515	4083	80	128		
16516	4084	06	6		LD B, +13
16517	4085	13	19		
16518	4086	D7	215	NEXT	RST 0010
16519	4087	10	16		DJNZ NEXT
16520	4088	FD	253		
16521	4089	C9	201		RET
16522	408A	3E	62	SQUARE	LD A, +11
16523	408B	11	17		
16524	408C	90	144		SUB B
16525	408D	47	71		LD B,A
16526	408E	CD	205		CALL PRINT AT
16527	408F	F5	245		
16528	4090	08	8		
16529	4091	3E	62		LD A, +80
16530	4092	80	128		
16531	4093	D7	215		RST 0010
16532	4094	C9	201		RET
16533	4095	CD	205	START	CALL SLOW
16534	4096	28	40		
16535	4097	0F	15		
16536	4098	CD	205		CALL CLS
16537	4099	2A	42		
16538	409A	0A	10		
16539	409B	CD	205		CALL LINE
16540	409C	82	130		
16541	409D	40	64		
16542	409E	06	6	EDGES	LD B, +10
16543	409F	10	16		
16544	40A0	C5	197	ROW	PUSH BC
16545	40A1	0E	14		LD C, +00
16546	40A2	00	0		
16547	40A3	CD	205		CALL SQUARE
16548	40A4	8A	138		
16549	40A5	40	64		
16550	40A6	C1	193		POP BC

16551	40A7	C5	197		PUSH BC
16552	40A8	0E	14		LD C, +12
16553	40A9	12	18		
16554	40AA	CD	205		CALL SQUARE
16555	40AB	8A	138		
16556	40AC	40	64		
16557	40AD	C1	193		POP BC
16558	40AE	10	16		DJNZ ROW
16559	40AF	F0	240		
16560	40B0	3E	62	NEW LINE	LD A, +75
16561	40B1	75	117		
16562	40B2	3C	60		INC A
16563	40B3	D7	215		RST 0010
16564	40B4	CD	205		CALL LINE
16565	40B5	82	130		
16566	40B6	40	64		
16567	40B7	C9	201		RET

THE SECOND MACHINE CODE ROUTINE

address. dec.	Hex.	Hex. code	Dec. code	Label	mnemonic
16568	40B8	–	–	UD	–
16569	40B9	–	–	LR	–
16570	40BA	–	–	X	–
16571	40BB	–	–	Y	–
16572	40BC	3E	62	VALUE	LD A, +01
16573	40BD	01	1		
16574	40BE	32	50		LD (UD),A
16575	40BF	B8	184		
16576	40C0	40	64		
16577	40C1	32	50		LD (LR),A
16578	40C2	B9	185		
16579	40C3	40	64		
16580	40C4	3C	60		INC A
16581	40C5	32	50		LD (X),A
16582	40C6	BA	186		
16583	40C7	40	64		
16584	40C8	ED	237	PLOT	LD BC,(X)
16585	40C9	4B	75		
16586	40CA	BA	186		
16587	40CB	40	64		
16588	40CC	3E	62		LD A, +9B

16589	40CD	9B	155		
16590	40CE	32	50		LD (T-ADDR),A
16591	40CF	30	48		
16592	40D0	40	64		
16593	40D1	CD	205		CALL PLOT c.r.
16594	40D2	B2	178		
16595	40D3	0B	11		
16596	40D4	CD	205	KEY TEST	CALL KEYBOARD
16597	40D5	BB	187		
16598	40D6	02	2		
16599	40D7	EB	235		EX DE,HL
16600	40D8	21	33		LD HL,+EFFB
16601	40D9	FB	251		
16602	40DA	EF	239		
16603	40DB	A7	167		AND A
16604	40DC	ED	237		SBC HL,DE
16605	40DD	52	82		
16606	40DE	C8	200		RET Z
16607	40DF	21	33		LD HL,+FD7F
16608	40E0	7F	127		
16609	40E1	FD	253		
16610	40E2	A7	167		AND A
16611	40E3	ED	237		SBC HL,DE
16612	40E4	52	82		
16613	40E5	C8	200		RET Z
16614	40E6	3A	58	X-TEST	LD A,(X)
16615	40E7	BA	186		
16616	40E8	40	64		
16617	40E9	FE	254		CP +02
16618	40EA	02	2		
16619	40EB	28	40		JR Z,LR-REV.
16620	40EC	04	4		
16621	40ED	FE	254		CP +23
16622	40EE	23	35		
16623	40EF	20	32		JR NZ,Y-TEST
16624	40F0	08	8		
16625	40F1	3A	58	LR REV	LD A.(LR)
16626	40F2	B9	185		
16627	40F3	40	64		
16628	40F4	2F	47		CPL
16629	40F5	3C	60		INC A
16630	40F6	32	50		LD (LR),A
16631	40F7	B9	185		

```
16632   40F8   40    64
16633   40F9   3A    58   Y-TEST     LD  A,(Y)
16634   40FA   BB   187
16635   40FB   40    64
16636   40FC   FE   254              CP  +0A
16637   40FD   0A    10
16638   40FE   28    40              JR  Z,UD-REV.
16639   40FF   04     4
16640   4100   FE   254              CP  +29
16641   4101   29    41
16642   4102   20    32              JR  NZ,UNPLOT
16643   4103   08     8
16644   4104   3A    58   UD REV     LD  A,(UD)
16645   4105   B8   184
16646   4106   40    64
16647   4107   2F    47              CPL
16648   4108   3C    60              INC  A
16649   4109   32    50              LD  (UD),A
16650   410A   B8   184
16651   410B   40    64
16652   410C   ED   237   UNPLOT     LD  BC,(X)
16653   410D   4B    75
16654   410E   BA   186
16655   410F   40    64
16656   4110   3E    62              LD A,+A0
16657   4111   A0   160
16658   4112   32    50              LD  (T-ADDR),A
16659   4113   30    48
16660   4114   40    64
16661   4115   CD   205              CALL PLOT c.r.
16662   4116   B2   178
16663   4117   0B    11
16664   4118   21    33   X-UPDATE   LD HL,+LR
16665   4119   B9   185
16666   411A   40    64
16667   411B   01     1              LD  BC,+X
16668   411C   BA   186
16669   411D   40    64
16670   411E   0A    10              LD  A,(BC)
16671   411F   96   150              SUB  (HL)
16672   4120   02     2              LD  (BC),A
16673   4121   2B    43   Y-UPDATE   DEC  HL
16674   4122   03     3              INC  BC
16675   4123   0A    10              LD  A,(BC)
```

16676	4124	96	150	SUB (HL)
16677	4125	02	2	LD (BC),A
16678	4126	18	24	JR PLOT
16679	4127	A0	160	

The table shows that the routine uses 166 locations (16514-16579 inclusively).

The final program can now be entered. Enter:

```
10 REM 1234567890 1234567890 123
4567890 1234567890 1234567890 12345
67890 1234567890 1234567890 1234567        166 locations
890 1234567890 1234567890 123456789        reserved.
0 1234567890 1234567890 12345678901
234567890 123456
20 LET K=USR 16533                         Location for START.
30 POKE 16571, INT (RND*32+10)             Set Y.
40 LET K=USR 16572                         VALUES.
50 RUN
```

Do not RUN this program until you have entered the machine code into line 10. SAVEing the program at this stage is advisable.

The machine code loader used on page 131 was a rather simple loader program but for this longer machine code REM statement a better loader program is desirable.

Enter the following lines:

```
200 LET A=0
210 FOR B=16514 TO 16679
220 INPUT C
230 POKE B,C
240 CLS
250 LET A=A+C
260 PRINT C,A
270 NEXT B
RUN 200
```

This loader program gives a 'checksum' which can be compared to the values given below and hence gives a good indication if the input-data is correct.

Enter:

	Checksum.
62,128,6,19,215,16,253,201,62,17,144,	1123
71,205,245,8,62,128,215,201,205,40,	2503
15,205,42,10,205,130,64,6,16,197,14,	3407
0,205,138,64,193,197,14,18,205,138,	4579

```
64,193,16,240,62,117,60,215,205,130,          5881
64,201,0,0,0,0,62,1,50,184,64,50,185          6742
64,60,50,186,64,237,75,186,64,62,155,         7945
50,48,64,205,178,11,205,187,2,235,33,         9163
251,239,167,237,82,200,33,127,253,           10752
167,237,82,200,58,186,64,254,2,40,4,         12046
254,35,32,8,58,185,64,47,60,50,185,          13024
64,58,187,64,254,10,40,4,254,41,32,          14032
8,58,184,64,47,60,50,184,64,237,75,          15063
186,64,62,160,50,48,64,205, 178,11,33        16124
185,64,1,186,64,10,150,2,43,3,10,150,        16992
2,24,160.                                    17178
```

Delete lines 200 - 270 and use SAVE "MOVING-BALL PROGRAM" before proceeding any further.

Now enter RUN. The moving ball should appear, and go round and round the playing area being reflected off each wall that it hits. Remember that the 'R' key is active and can be used to restart the ball. Also the 'BREAK' key is active.

The reader might like to add lines to the program so as to create 'targets'.

e.g.
```
35 PLOT 20,20
36 PLOT 22,22
37 PLOT 25,19
RUN
```

The reader might also like to increase the size of the playing area by changing the appropriate values in the machine code program.

7.5 And So On . . .

The machine code routine for the first example was only 17 bytes in length, whereas the two routines for the second example together used 166 locations, and the 8K monitor program uses 8192!

It is hoped that the reader can now appreciate just how a large machine code program is written.

The two example programs in this chapter are really only an introduction to the techniques of machine code programming but by building upon the suggested techniques it is possible to write very impressive programs.

Good Luck.

Appendix i.

The SAVE command routine.

address	Hex. code	mnemonic		comment
02F6	CD A8 03	CALL	03A8	Check a name is given.
2F9	38 F9	JR	C,02F4	Jump if no name - error 9.
2FB	EB	EX	DE,HL	
2FC	11 CB 12	LD	DE,+12CB	5 second timer.
2FF	CD 43 0F	CALL	0F43	Test for Break Key.
302	30 2E	JR	NC,0332	Break pressed
304	10 FE	DJNZ	0304	Delay loop.
306	1B	DEC	DE	Delay loop.
307	7A	LD	A,D	
308	B3	OR	E	
309	20 F4	JR	NZ,02FF	End of delay?
30B	CD 1E 03	CALL	031E	SAVE a byte of the name.
30E	CB 7E	BIT	7,(HL)	Last byte of name?
310	23	INC	HL	Next byte.
311	28 F8	JR	Z,030B	Name done?
313	21 09 40	LD	HL,+4009	Start sending system variables
316	CD 1E 03	CALL	031E	Each byte.
319	CD FC 01	CALL	01FC	Update routine (see below).
31C	18 F8	JR	0316	Back for next byte.
31E	5E	LD	E,(HL)	Byte into E.
31F	37	SCF		Set the 'marker'.
320	CB 13	RL	E	Rotate 'marker' in.
322	C8	RET	Z	8 bits done?
323	9F	SBC	A,A	A=00 or A=FF.
324	E6 05	AND	+05	A=00 or A=05.
326	C6 04	ADD	+04	A=04 or A=09.
328	4F	LD	C,A	Save in C.
329	D3 FF	OUT	(+FF),A	To the cassette player.
32B	06 23	LD	B,+23	Timing loop.
32D	10 FE	DJNZ	032D	
32F	CD 43 0F	CALL	0F43	Test for Break Key.
332	30 72	JR	NC,03A6	Error D if Break pressed.
334	06 1E	LD	B,+1E	Timing Loop.
336	10 FE	DJNZ	0336	
338	0D	DEC	C	Decrease counter.
339	20 EE	JR	NZ,0329	Bit finished?

33B	A7	AND	A	Timing loop.
33C	10 FD	DJNZ	033B	
33E	18 E0	JR	0320	Next bit.
01FC	23	INC	HL	Next byte.
1FD	EB	EX	DE,HL	Exchange.
1FE	2A 14 40	LD	HL,(4014)	All bytes up to E-Line.
201	37	SCF		
202	ED 52	SBC	HL,DE	Test for end.
204	EB	EX	DE,HL	
205	D0	RET	NC	(Return to SAVE next byte.
206	E1	POP	HL	(SAVEing finished so
207...				(continue with display.
				(or
				(Return to LOAD next byte.
				(LOADing finished so
				(continue with display.

The LOAD command routine.

address	Hex. code	mnemonic		comment
0340	CD A8 03	CALL	03A8	Check if name is given.
343	CB 12	RL	D	
345	CB 0A	RRC	D	
347	CD 4C 03	CALL	034C	Start listening to tape.
34A	18 FB	JR	0347	
34C	0E 01	LD	C,+01	
34E	06 00	LD	B,+00	
350	3E 7F	LD	A,+FF	Test Break Key.
352	DB FE	IN	A,(+FE)	
354	D3 FF	OUT	(+FF),A	The 'echo' to the screen.
356	1F	RRA		
357	30 49	JR	NC,03A2	Jump if Break pressed.
359	17	RLA		
35A	17	RLA		
35B	38 28	JR	C,0385	Build-up a byte in C, if C is set.
35D	10 F1	DJNZ	0350	
35F	F1	POP	AF	
360	BA	CP	D	
361	D2 E5 03	JP	NC,03E5	Re-initialise if needed.
364	62	LD	H,D	Transfer the 'address of the
365	6B	LD	L,E	program name' to HL.
366	CD 4C 03	CALL	034C	Start listening.

address	Hex. code	mnemonic		comment
369	CB 7A	BIT	7,D	
36B	79	LD	A,C	
36C	2Ø Ø3	JR	NZ,Ø371	
36E	BE	CP	(HL)	Match letters of 'name'.
36F	2Ø D6	JR	NZ,Ø347	
371	23	INC	HL	Next letter of name.
372	17	RLA		
373	3Ø F1	JR	NC,Ø366	
375	FD 34 15	INC	(4Ø15)	
378	21 Ø9 4Ø	LD	HL,4ØØ9	Start LOADing at 4ØØ9.
37B	5Ø	LD	D,B	
37C	CD 4C Ø3	CALL	Ø34C	Start listening to tape.
37F	71	LD	(HL),C	Put byte in RAM.
380	CD FC Ø1	CALL	Ø1FC	Update routine (see SAVE).
383	18 F6	JR	Ø37B	Next byte.
385	D5	PUSH	DE	
386	1E 94	LD	E,+94	Timing loops.
388	Ø6 1A	LD	B,+1A	
38A	1D	DEC	E	
38B	DB FE	IN	A,(+FE)	Current signal on tape.
38D	17	RLA		
38E	CB 7B	BIT	7,E	
Ø39Ø	7B	LD	A,E	Build-up
391	38 F5	JR	C,Ø388	
393	1Ø F5	DJNZ	Ø38A	Byte in
395	D1	POP	DE	
396	2Ø Ø4	JR	NZ,Ø39C	C register.
398	FE 56	CP	+56	
39A	3Ø B2	JR	NC,Ø34E	
39C	3F	CCF		
39D	CB 11	RL	C	
39F	3Ø AD	JR	NC,Ø34E	
3A1	C9	RET		Byte finished so return to Ø369 or Ø37F.
Ø3A2	7A	LD	A,D	Re-initialise if
3A3	A7	AND	A	D is zero.
3A4	28 BB	JR	Z,Ø361	otherwise error 'D'.
Ø3A6	CF	RST	ØØØ8	Error 'D' – Break key pressed.
3A7	ØC	'ØC'		

Keyboard Scanning Routine

address	Hex. code	mnemonic		comment
Ø2BB	21 FF FF	LD	HL,+FFFF	Initialise HL.
2BE	Ø1 FE FE	LD	BC,+FEFE	The first port address.
2C1	ED 78	IN	A,(C)	Read the first line.
2C3	F6 Ø1	OR	+Ø1	Ignore Bit Ø.
2C5	F6 EØ	OR	+EØ	Ignore Bits 5,6,7.
2C7	57	LD	D,A	Build-up
2C8	2F	CPL		the
2C9	FE Ø1	CP	+Ø1	Key Value
2CB	9F	SBC	A,A	in
2CC	BØ	OR	B	HL.
2CD	A5	AND	L	
2CE	6F	LD	L,A	
2CF	7C	LD	A,H	
2DØ	A2	AND	D	
2D1	67	LD	H,A	
2D2	CB ØØ	RLC	B	
2D4	ED 78	IN	A,(C)	Read the other lines.
2D6	38 ED	JR	C,Ø2C5	8 reads done?
2D8	1F	RRA		
2D9	CB 14	RL	H	Final key value in HL.
2DB	17	RLA		
2DC	17	RLA		Test
2DD	17	RLA		line 6
2DE	9F	SBC	A,A	UK/USA
2DF	E6 18	AND	+18	standard
2E1	C6 1F	ADD	+1F	ZX-81.
2E3	32 28 4Ø	LD	(4Ø28),A	
2E6	C9	RET		Return.

(see also appendix iv. 'The Table of Key Values'.) page

Keyboard decode routine

address	Hex. code	mnemonic		comment
Ø7BD	16 ØØ	LD	D,+ØØ	Enter with Key Value in BC.
7BF	CB 28	SRA	B	
7C1	9F	SBC	A,A	Manipulate
7C2	F6 26	OR	+26	the Key values.
7C4	2E Ø5	LD	L,+Ø5	to produce
7C6	95	SUB	L	the A register
7C7	85	ADD	A,L	holding
7C8	37	SCF		1-78 (decimal).

```
7C9    CB 19      RR    C
7CB    38 FA      JR    C,Ø7C7
7CD    ØC         INC   C
7CE    CØ         RET   NZ
7CF    48         LD    C,B
7DØ    2D         DEC   L
7D1    23 Ø1      LD    L,+Ø1
7D3    2Ø F2      JR    NZ,+Ø7C7
7D5    21 7D ØØ   LD    HL,+ØØ7D   Base address of character
                                   table.
7D8    5F         LD    E,A        Transfer to E.
7D9    19         ADD   HL,DE      Form the address in HL.
7DA    37         SCF
7DB    C9         RET              Return with address of the
                                   character code in HL.
```

Appendix ii.

Tables of Z80 Machine Code Language Instructions.

00 NOP	01 LD BC, +dddd	02 LD (BC),A	03 INC BC	04 INC B	05 DEC B	06 LD B,+dd	07 RLCA
10 DJNZ e	11 LD DE, +dddd	12 LD (DE),A	13 INC DE	14 INC D	15 DEC D	16 LD D,+dd	17 RLA
20 JR NZ,e	21 LD HL, +dddd	22 LD (addr.) HL	23 INC HL	24 INC H	25 DEC H	26 LD H,+dd	27 DAA
30 JR NC,e	31 LD SP, +dddd	32 LD (addr.), A	33 INC SP	34 INC (HL)	35 DEC (HL)	36 LD (HL), +dd	37 SCF
40 LD B,B	41 LD B,C	42 LD B,D	43 LD B,E	44 LD B,H	45 LD B,L	46 LD B,(HL)	47 LD B,A
50 LD D,B	51 LD D,C	52 LD D,D	53 LD D,E	54 LD D,H	55 LD D,L	56 LD D,(HL)	57 LD D,A
60 LD H,B	61 LD H,D	62 LD H,C	63 LD H,E	64 LD H,H	65 LD H,L	66 LD H,(HL)	67 LD H,A
70 LD (HL),B	71 LD (HL),C	72 LD (HL),D	73 LD (HL),E	74 LD (HL),H	75 LD (HL),L	76 HALT	77 LD (HL),A
80 ADD A,B	81 ADD A,C	82 ADD A,D	83 ADD A,E	84 ADD A,H	85 ADD A,L	86 ADD A,(HL)	87 ADD A,A
90 SUB B	91 SUB C	92 SUB D	93 SUB E	94 SUB H	95 SUB L	96 SUB (HL)	97 SUB A

AØ AND B	A1 AND C	A2 AND D	A3 AND E	A4 AND H	A5 AND L	A6 AND (HL)	A7 AND A
BØ OR B	B1 OR C	B2 OR D	B3 OR E	B4 OR H	B5 OR L	B6 OR (HL)	B7 OR A
CØ RET NZ	C1 POP BC	C2 JP NZ,addr.	C3 JP addr.	C4 CALL NZ,addr.	C5 PUSH BC	C6 ADD A,+dd	C7 RST ØØØØ
DØ RET NC	D1 POP DE	D2 JP NC,addr.	D3 OUT (+dd) A	D4 CALL NC,addr.	D5 PUSH DE	D6 SUB +dd	D7 RST ØØ1Ø
EØ RET PO	E1 POP HL	E2 JP PO,addr	E3 EX (HL),SP	E4 CALL PO,addr	E5 PUSH HL	E6 AND +dd	E7 RST ØØ2Ø
FØ RET P	F1 POP AF	F2 JP P,addr.	F3 DI	F4 CALL P,addr.	F5 PUSH AF	F6 OR +dd	F7 RST ØØ3Ø
Ø8 EX AF,A'F'	Ø9 ADD HL,BC	ØA LD A,(BC)	ØB DEC BC	ØC INC C	ØD DEC C	ØE' LD C,+dd	ØF RRCA
18 JR e	19 ADD HL,DE	1A LD A,(DE)	1B DEC DE	1C INC E	1D DEC E	1E LD E,+dd	1F RRA
28 JR Z,e	29 ADD HL,HL	2A LD HL,(addr.)	2B DEC HL	2C INC L	2D DEC L	2E LD L,+dd	2F CPL
38 JR C,e	39 ADD HL,SP	3A LD A,(addr.)	3B DEC SP	3C INC A	3D DEC A	3E LD A,+dd	3F CCF
48 LD C,B	49 LD C,C	4A LD C,D	4B LD C,E	4C LD C,H	4D LD C,L	4E LD C,(HL)	4F LD C,A
58 LD E,B	59 LD E,C	5A LD E,D	5B LD E,E	5C LD E,H	5D LD E,L	5E LD E,(HL)	5F LD E,A

68 LD L,B	69 LD L,C	6A LD L,D	6B LD L,E	6C LD L,H	6D LD L,L	6E LD L,(HL)	6F LD L,A
78 LD A,B	79 LD A,C	7A LD A,D	7B LD A,E	7C LD A,H	7D LD A,L	7E LD A,(HL)	7F LD A,A
88 ADC A,B	89 ADC A,C	8A ADC A,D	8B ADC A,E	8C ADC A,H	8D ADC A,L	8E ADC A,(HL)	8F ADC A,A
98 SBC A,B	99 SBC A,C	9A SBC A,D	9B SBC A,E	9C SBC A,H	9D SBC A,L	9E SBC A,(HL)	9F SBC A,A
A8 XOR B	A9 XOR C	AA XOR D	AB XOR E	AC XOR H	AD XOR L	AE XOR (HL)	AF XORA
B8 CP B	B9 CP C	BA CP D	BB CP E	BC CP H	BD CP L	BE CP (HL)	BF CP A
C8 RET Z	C9 RET	CA JP Z,addr.	CB see pages 68, 70	CC CALL Z,addr.	CD CALL addr.	CE ADC A,+dd	CF RST 0008
D8 RET C	D9 EXX	DA JP C,addr.	DB IN A,(+dd)	DC CALL C,addr.	DD see page 158	DE SBC A,+dd	DF RST 0018
E8 RET PE	E9 JP (HL)	EA JP PE,addr.	EB EX DE,HL	EC CALL PE,addr.	ED see page 158	EE XOR +dd	EF RST 0028
F8 RET M	F9 LD SP,HL	FA JP M,addr.	FB EI	FC CALL M,addr.	FD see page 158	FE CP +dd	FF RST 0038

ED instructions

ED 4Ø IN B,(C)	ED 5Ø IN D,(C)	ED 6Ø IN H,(C)		ED AØ LDI	ED BØ LDIR
ED 41 OUT (C),B	ED 51 OUT (C),D	ED 61 OUT (C),H		ED A1 CPI	ED B1 CPIR
ED 42 SBC HL,BC	ED 52 SBC HL,DE	ED 62 SBC HL,HL	ED 72 SBC HL,SP	ED A2 INI	ED B2 INIR
ED 43 LD (addr.),BC	ED 53 LD (addr.),DE	ED 63 LD (addr.),HL	ED 73 LD (addr.),SP	ED A3 OUTI	ED B3 OTIR
ED 44 NEG					
ED 45 RETN					
ED 46 IM Ø	ED 56 IM 1	ED 66 IM 2			
ED 47 LD I,A	ED 57 LD A,I	ED 67 RRD			
ED 48 IN C,(C)	ED 58 IN E,(C)	ED 68 IN L,(C)	ED 78 IN A,(C)	ED A8 LDD	ED B8 LDDR
ED 49 OUT (C),C	ED 59 OUT (C),E	ED 69 OUT (C),L	ED 79 OUT (C),A	ED A9 CPD	ED B9 CPDR
ED 4A ADC HL,BC	ED 5A ADC HL,DE	ED 6A ADC HL,HL	ED 7A ADC HL,SP	ED AA IND	ED BA INDR
ED 4B LD BC,(addr.)	ED 5B LD DE,(addr.)	ED 6B LD HL,(addr.)	ED 7B LD SP,(addr.)	ED AB OUTD	ED BB OTRD
ED 4D RETI					
ED 4F LD R,A	ED 5F LD A,R	ED 6F RLD			

The Indexing Instructions

All the instructions using the IX register pair are prefixed by 'DD'. All the instructions using the IY register pair are prefixed by 'FD'. In the following simply read IY for IX and change DD to FD when dealing with the IY register pair.

DD 09		ADD IX,BC	DD CB d	06	RLC	(IX+d)
DD 19		ADD IX,DE	DD CB d	OE	RRC	(IX+d)
DD 21	+dddd	LD IX,+dddd	DD CB d	16	RL	(IX+d)
DD 22	addr.	LD (addr.),IX	DD CB d	1E	RR	(IX+d)
DD 23		INC IX	DD CB d	26	SLA	(IX+d)
DD 29		ADD IX,IX	DD CB d	2E	SRA	(IX+d)
DD 2A	addr.	LD IX,(addr.)	DD CB d	3E	SRL	(IX+d)
DD 2B		DEC IX	DD CB d	46	BIT	0,(IX+d)
DD 34	d	INC (IX+d)	DD CB d	4E	BIT	1,(IX+d)
DD 35	d	DEC (IX+d)	DD CB d	56	BIT	2,(IX+d)
DD 36	d+dd	LD (IX+d),+dd	DD CB d	5E	BIT	3,(IX+d)
DD 39		ADD IX,SP	DD CB d	66	BIT	4,(IX+d)
DD 46	d	LD B,(IX+d)	DD CB d	6E	BIT	5,(IX+d)
DD 4E	d	LD C,(IX+d)	DD CB d	76	BIT	6,(IX+d)
DD 56	d	LD D,(IX+d)	DD CB d	7E	BIT	7,(IX+d)
DD 5E	d	LD E,(IX+d)	DD CB d	86	RES	0,(IX+d)
DD 66	d	LD H,(IX+d)	DD CB d	8E	RES	1,(IX+d)
DD 6E	d	LD L,(IX+d)	DD CB d	96	RES	2,(IX+d)
DD 70	d	LD (IX+d),B	DD CB d	9E	RES	3,(IX+d)
DD 71	d	LD (IX+d),C	DD CB d	A6	RES	4,(IX+d)
DD 72	d	LD (IX+d),D	DD CB d	AE	RES	5,(IX+d)
DD 73	d	LD (IX+d),E	DD CB d	B6	RES	6,(IX+d)
DD 74	d	LD (IX+d),H	DD CB d	BE	RES	7,(IX+d)
DD 75	d	LD (IX+d),L	DD CB d	C6	SET	0,(IX+d)
DD 77	d	LD (IX+d),A	DD CB d	CE	SET	1,(IX+d)
DD 7E	d	LD A,(IX+d)	DD CB d	D6	SET	2,(IX+d)
DD 86	d	ADD A,(IX+d)	DD CB d	DE	SET	3,(IX+d)
DD 8E	d	ADC A,(IX+d)	DD CB d	E6	SET	4,(IX+d)
DD 96	d	SUB (IX+d)	DD CB d	EE	SET	5,(IX+d)
DD 9E	d	SBC A,(IX+d)	DD CB d	F6	SET	6,(IX+d)
DD A6	d	AND (IX+d)	DD CB d	FE	SET	7,(IX+d)
DD AE	d	XOR (IX+d)	DD E1		POP	IX
DD B6	d	OR (IX+d)	DD E3		EX	(SP),IX
DD BE	d	CP (IX+d)	DD E5		PUSH	IX
			DD E9		JP	(IX)
			DD F9		LD	SP,IX

Appendix iii.

Decimal-Hexadecimal Conversion Table

DECIMAL 0-255						HEX. 00-FF LOW BYTE			
DEC.	HEX.	DEC.	HEX.	DEC.	HEX.	2's C.	DEC.	HEX.	2's C.
0	00	64	40	128	80	-128	192	C0	-64
1	01	65	41	129	81	-127	193	C1	-63
2	02	66	42	130	82	-126	194	C2	-62
3	03	67	43	131	83	-125	195	C3	-61
4	04	68	44	132	84	-124	196	C4	-60
5	05	69	45	133	85	-123	197	C5	-59
6	06	70	46	134	86	-122	198	C6	-58
7	07	71	47	135	87	-121	199	C7	-57
8	08	72	48	136	88	-120	200	C8	-56
9	09	73	49	137	89	-119	201	C9	-55
10	0A	74	4A	138	8A	-118	202	CA	-54
11	0B	75	4B	139	8B	-117	203	CB	-53
12	0C	76	4C	140	8C	-116	204	CC	-52
13	0D	77	4D	141	8D	-115	205	CD	-51
14	0E	78	4E	142	8E	-114	206	CE	-50
15	0F	79	4F	143	8F	-113	207	CF	-49
16	10	80	50	144	90	-112	208	D0	-48
17	11	81	51	145	91	-111	209	D1	-47
18	12	82	52	146	92	-110	210	D2	-46
19	13	83	53	147	93	-109	211	D3	-45
20	14	84	54	148	94	-108	212	D4	-44
21	15	85	55	149	95	-107	213	D5	-43
22	16	86	56	150	96	-106	214	D6	-42
23	17	87	57	151	97	-105	215	D7	-41
24	18	88	58	152	98	-104	216	D8	-40
25	19	89	59	153	99	-103	217	D9	-39
26	1A	90	5A	154	9A	-102	218	DA	-38
27	1B	91	5B	155	9B	-101	219	DB	-37
28	1C	92	5C	156	9C	-100	220	DC	-36
29	1D	93	5D	157	9D	- 99	221	DD	-35
30	1E	94	5E	158	9E	- 98	222	DE	-34
31	1F	95	5F	159	9F	- 97	223	DF	-33
32	20	96	60	160	A0	- 96	224	E0	-32
33	21	97	61	161	A1	- 95	225	E1	-31
34	22	98	62	162	A2	- 94	226	E2	-30
35	23	99	63	163	A3	- 93	227	E3	-29
36	24	100	64	164	A4	- 92	228	E4	-28
37	25	101	65	165	A5	- 91	229	E5	-27
38	26	102	66	166	A6	- 90	230	E6	-26
39	27	103	67	167	A7	- 89	231	E7	-25
40	28	104	68	168	A8	- 88	232	E8	-24
41	29	105	69	169	A9	- 87	233	E9	-23
42	2A	106	6A	170	AA	- 86	234	EA	-22
43	2B	107	6B	171	AB	- 85	235	EB	-21
44	2C	108	6C	172	AC	- 84	236	EC	-20
45	2D	109	6D	173	AD	- 83	237	ED	-19
46	2E	110	6E	174	AE	- 82	238	EE	-18
47	2F	111	6F	175	AF	- 81	239	EF	-17
48	30	112	70	176	B0	- 80	240	F0	-16
49	31	113	71	177	B1	- 79	241	F1	-15
50	32	114	72	178	B2	- 78	242	F2	-14
51	33	115	73	179	B3	- 77	243	F3	-13
52	34	116	74	180	B4	- 76	244	F4	-12
53	35	117	75	181	B5	- 75	245	F5	-11
54	36	118	76	182	B6	- 74	246	F6	-10
55	37	119	77	183	B7	- 73	247	F7	- 9
56	38	120	78	184	B8	- 72	248	F8	- 8
57	39	121	79	185	B9	- 71	249	F9	- 7
58	3A	122	7A	186	BA	- 70	250	FA	- 6
59	3B	123	7B	187	BB	- 69	251	FB	- 5
60	3C	124	7C	188	BC	- 68	252	FC	- 4
61	3D	125	7D	189	BD	- 67	253	FD	- 3
62	3E	126	7E	190	BE	- 66	254	FE	- 2
63	3F	127	7F	191	BF	- 65	255	FF	- 1

Decimal-Hexadecimal Conversion Table (cont.)

DECIMAL 0-65280 HEX. 00-FF, HIGH BYTE

DECIMAL	HEX.	DECIMAL	HEX.	DECIMAL	HEX.	DECIMAL	HEX.
0	00	16384	40	32768	80	49152	C0
256	01	16640	41	33024	81	49408	C1
512	02	16896	42	33280	82	49664	C2
768	03	17152	43	33536	83	49920	C3
1024	04	17408	44	33792	84	50176	C4
1280	05	17664	45	34048	85	50432	C5
1536	06	17920	46	34304	86	50688	C6
1792	07	18176	47	34560	87	50944	C7
2048	08	18432	48	34816	88	51200	C8
2304	09	18688	49	35072	89	51456	C9
2560	0A	18944	4A	35328	8A	51712	CA
2816	0B	19200	4B	35584	8B	51968	CB
3072	0C	19456	4C	35840	8C	52224	CC
3328	0D	19712	4D	36096	8D	52480	CD
3584	0E	19968	4E	36352	8E	52736	CE
3840	0F	20224	4F	36608	8F	52992	CF
4096	10	20480	50	36864	90	53248	D0
4352	11	20736	51	37120	91	53504	D1
4608	12	20992	52	37376	92	53760	D2
4864	13	21248	53	37632	93	54016	D3
5120	14	21504	54	37888	94	54272	D4
5376	15	21760	55	38144	95	54528	D5
5632	16	22016	56	38400	96	54784	D6
5888	17	22272	57	38656	97	55040	D7
6144	18	22528	58	38912	98	55296	D8
6400	19	22784	59	39168	99	55552	D9
6656	1A	23040	5A	39424	9A	55808	DA
6912	1B	23296	5B	39680	9B	56064	DB
7168	1C	23552	5C	39936	9C	56320	DC
7424	1D	23808	5D	40192	9D	56576	DD
7680	1E	24064	5E	40448	9E	56832	DE
7936	1F	24320	5F	40704	9F	57088	DF
8192	20	24576	60	40960	A0	57344	E0
8448	21	24832	61	41216	A1	57600	E1
8704	22	25088	62	41472	A2	57856	E2
8960	23	25344	63	41728	A3	58112	E3
9216	24	25600	64	41984	A4	58368	E4
9472	25	25856	65	42240	A5	58624	E5
9728	26	26112	66	42496	A6	58880	E6
9984	27	26368	67	42752	A7	59136	E7
10240	28	26624	68	43008	A8	59392	E8
10496	29	26880	69	43264	A9	59648	E9
10752	2A	27136	6A	43520	AA	59904	EA
11008	2B	27392	6B	43776	AB	60160	EB
11264	2C	27648	6C	44032	AC	60416	EC
11520	2D	27904	6D	44288	AD	60672	ED
11776	2E	28160	6E	44544	AE	60928	EE
12032	2F	28416	6F	44800	AF	61184	EF
12288	30	28672	70	45056	B0	61440	F0
12544	31	28928	71	45312	B1	61696	F1
12800	32	29184	72	45568	B2	61952	F2
13056	33	29440	73	45824	B3	62208	F3
13312	34	29696	74	46080	B4	62464	F4
13568	35	29952	75	46336	B5	62720	F5
13824	36	30208	76	46592	B6	62976	F6
14080	37	30464	77	46848	B7	63232	F7
14336	38	30720	78	47104	B8	63488	F8
14592	39	30976	79	47360	B9	63744	F9
14848	3A	31232	7A	47616	BA	64000	FA
15104	3B	31488	7B	47872	BB	64256	FB
15360	3C	31744	7C	48128	BC	64512	FC
15616	3D	32000	7D	48384	BD	64768	FD
15872	3E	32256	7E	48640	BE	65024	FE
16128	3F	32512	7F	48896	BF	65280	FF

Appendix iv.
Table of 'Key Values'

Key	Hex. Value	Dec. Value	Key	Hex. Value	Dec. Value
1	FDF7	65015	EDIT	FCF7	64759
2	FBF7	64503	AND	FAF7	64247
3	F7F7	63479	THEN	F6F7	63223
4	EFF7	61431	TO	EEF7	61175
5	DFF7	57335	←	DEF7	57079
6	DFEF	57327	↓	DEEF	57071
7	EFEF	61423	↑	EEEF	61167
8	F7EF	63471	→	F6EF	63215
9	FBEF	64495	GRAPHICS	FAEF	64239
0	FDEF	65007	RUBOUT	FCEF	64751
Q	FDFB	65019	""	FCFB	64763
W	FBFB	64507	OR	FAFB	64251
E	F7FB	63438	STEP	F6FB	63227
R	EFFB	61435	<=	EEFB	61179
T	DFFB	57339	<>	DEFB	57083
Y	DFDF	57311	>=	DEDF	57055
U	EFDF	61407	S	EEDF	61151
I	F7DF	63455	(	F6DF	63199
O	FBDF	64479	)	FADF	64223
P	FDDF	64991	"	FCDF	64735
A	FDFD	65021	STOP	FCFD	64765
S	FBFD	64509	LPRINT	FAFD	64253
D	F7FD	63485	SLOW	F6FD	63229
F	EFFD	61437	FAST	EEFD	61181
G	DFFD	57341	LLIST	DEFD	57085
H	DFBF	57279	**	DEBF	57023
J	EFBF	61375	–	EEBF	61119
K	F7BF	63423	+	F6BF	63167
L	FBBF	64447	=	FABF	64191
NEWLINE	FDBF	64959	FUNCTION	FCBF	64959
Z	FBFE	64510	:	FAFE	64254
X	F7FE	63486	;	F6FE	63230
C	EFFE	61438	?	EEFE	61182
V	DFFE	57342	/	DEFE	57086
B	DF7F	57215	*	DE7F	56959
N	EF7F	61311	<	EE7F	61055
M	F77F	63359	>	F67F	63103
•	FB7F	64383	,	FA7F	64127
SPACE	FD7F	64895	£	FC7F	64639
NO KEY	FFFF	65535			

You may also enjoy...

You may also enjoy...

Made in the USA
Monee, IL
08 November 2026

68002054R00100